FOREST BIOMASS

FORESTRY SCIENCES

Also in this series:

Prins CFL ed: Production, Marketing and Use of Finger-Jointed Sawnwood.
ISBN 90-247-2569-0
Oldeman RAA, et al. eds: Tropical Hardwood Utilization: Practice and
Prospects. 1982. ISBN 90-247-2581-X
Baas P ed: New Perspectives in Wood Anatomy, 1982. ISBN 90-247-2526-7

In preparation:

Bonga JM and Durzan DJ: Tissue Culture in Forestry. 1982 ISBN 90-247-
2660-3
Chandler CC, Cheney P and Williams DF, eds: Fire in Forest
Den Ouden P and Boom BK, eds: Manual of Cultivated Conifers: Hardy in
Cold- and Warm-Temperate Zone. 1982. ISBN 90-247-2148-2
Gordon JC and Wheeler CT eds: Biological Nitrogen Fixation in Forest
Ecosystems: Foundation and Applications
Hummel FC ed: Forestry Policy
Németh MV: The Virus — Mycoplasma and Rickettsia Disease of Fruit Trees
Powers RF and Miller HG eds: Applied Aspects of Forest Tree Nutrition
Powers RF and Miller HG eds: Basic Aspects of Forest Tree Nutrition
Rajagopal R: Information Analysis for Resource Management
Van Nao T, ECE/FAO/Agriculture and Timber Division ed: Forest Fire
Prevention and Control

FOREST BIOMASS

written and translated by

T. SATOO

edited and revised by

H.A.I. MADGWICK

1982

MARTINUS NIJHOFF / DR W. JUNK PUBLISHERS
THE HAGUE / BOSTON / LONDON

Distributors

for the United States and Canada
Kluwer Boston, Inc.
190 Old Derby Street
Hingham, MA 02043
USA

for all other countries
Kluwer Academic Publishers Group
Distribution Center
P.O. Box 322
3300 AH Dordrecht
The Netherlands

Library of Congress Cataloging in Publication Data

Satoo, Taishichirō, 1917-
   Forest biomass.

   (Forestry sciences)
   Revised translation of: Rikujō shokubutsu
gunraku no busshitsu seisan. Shinrin.
   Bibliography: p.
   Includes index.
   1. Forest ecology. 2. Primary productivity
(Biology) I. Madgwick, H.A.I. II. Title.
III. Series.
QK938.F6S2313 1982   574.5'2642   82-8074
                                   AACR2

ISBN-13: 978-94-009-7629-0   e-ISBN-13: 978-94-009-7627-6
DOI: 10.1007/ 978-94-009-7627-6

CONTENTS

                                                        Page

Preface to the English edition   vii

Chapter

1   Primary Production   1

2   Forests   5

3   Methods of Estimating Forest Biomass   15

4   Biomass   46

5   Production   90

6   Factors Affecting Rates of Production   119

    References   135

    Index   151

PREFACE TO THE ENGLISH EDITION

Lord Rutherford has said that all science is either physics or stamp collecting. On that basis the study of forest biomass must be classified with stamp collecting and other such pleasurable pursuits. Japanese scientists have led the world, not only in collecting basic data, but in their attempts to systematise our knowledge of forest biomass. They have studied factors affecting dry matter production of forest trees in an attempt to approach underlying physical principles. This edition of Professor Satoo's book has been made possible the help of Dr John F. Hosner and the Virginia Polytechnical Institute and State University who invited Dr Satoo to Blacksburg for three months in 1973 at about the time when he was in the final stages of preparing the Japanese version. Since then the explosion of world literature on forest biomass has continued to be fired by increasing shortages of timber supplies in many parts of the world as well as by a need to explore renewable sources of energy. In revising the original text I have attempted to maintain the input of Japanese work - much of which is not widely available outside Japan - and to update both the basic information and, where necessary, the conclusions to keep them in tune with current thinking. Those familiar with the Japanese original will find Chapter 3 largely rewritten on the basis of new work - much of which was initiated while Dr Satoo was in Blacksburg. Major additions have been made to Chapter 4. For these and all other amendments I accept full responsibility.

Few, apart from other authors, will appreciate the debt of gratitude that I feel to the technical staff at the Forest Research Institute in Rotorua for their efforts to assist in the preparation of this work. I am particularly grateful to the staff in the draughting, photographic, typing and editorial sections for their unstinted efforts. My thanks are also due to Dr J.D. Ovington who first

VIII

introduced me to the study of forest biomass and encouraged me to pursue the subject by further study. I am indebted to those many scientists who have contributed to my current knowledge and whose ideas I have been able to consciously or subconsciously use in this book. Finally I wish to thank my family for the sacrifices they have made in the past months during the preparation of this manuscript.

Rotorua   H.A.I. Madgwick
  March 1982

1. PRIMARY PRODUCTION

For survival and growth, living organisms must obtain energy and minerals from their environment and must synthesize organic matter. The synthesis of organic matter through photosynthesis by green plants in an ecosystem is called the primary production of that ecosystem. The total amount of organic matter produced by photosynthesis is called gross production (Pg). Green plants consume some photosynthate in respiration (r), the remainder being incorporated into the body of the plant. This we call net production (Pn). So,

$$Pg = Pn + r \quad \dots\dots\dots\dots\dots\dots\dots\dots\dots\dots\dots \quad (1.1)$$

$$\text{or} \quad Pn = Pg - r \quad \dots\dots\dots\dots\dots\dots\dots\dots\dots\dots\dots \quad (1.2)$$

The values of Pg and Pn are usually expressed as dry weight in which case they are often called dry matter production. For forest eco- systems it is general practice to use years and hectares as the units of time and area, respectively. We can either determine net production by measuring gross production and respiration or we can determine gross production by measuring net production and respiration.

There are many ways to estimate gross production. One group of methods relies on the estimation of carbon dioxide exchange through photosynthesis over time (Monsi and Saeki, 1953; Lange and Schultze, 1971). Alternatively, the resultant plant material can be cut and weighed using harvesting techniques to determine net production with an allowance for respiration as in equation (1.1). The harvest method has been widely used throughout the world and most of the information in this book is based on this method.

1. OBJECTIVES OF THE STUDIES OF PRIMARY PRODUCTION

In the last thirty years, the production in forest ecosystems has been widely studied from a variety of viewpoints. Some researchers take a geographic standpoint, relating the distribution of productivity over the earth's surface to either climatic factors (Paterson 1956 Lieth 1972) or the distribution of the main plant communities (Rodin and Bazilevich 1965). In this approach we get a rather general idea of productivity over the earth while neglecting the productive relationships in any particular forest stand. Other studies concentrate on the pattern and mechanism of production within a given forest stand as a basis for understanding the energy flow and circulation of nutrients in the forest as an ecosystem. Such an understanding of the functioning of forest ecosystems is a prerequisite to their wise use as a source of renewable raw materials. As a consequence biomass estimation is widely used in conjunction with studies of forest management such as fertilisation and thinning.

The oil crises of the 1970s have led a number of investigators to study the productivity of forests as a source of energy and chemical feed stock. Such studies have frequently been related to attempts to define the potential nutrient drain from forests under a wide range of harvesting practices including conventional clearcutting, 'whole-tree harvesting' in which all the above-ground parts of the trees are harvested, and 'complete-tree harvesting' involving the harvesting of roots as well.

2. A BRIEF HISTORY OF EARLIER STUDIES

It is only during the last two decades that studies of primary production of forest ecosystems have been made worldwide and by many scientists. As Walter (1951) pointed out, the interest of biologists had been concentrated on the physiological processes of plants and not directed to the entire process of production of organic matter in ecosystems. On the other hand, agronomists and forest scientists had tended to study the yield of what could be harvested and neglected the production of organic matter as the basic process determining yield. However, as has happened in many other fields of science, there were some early pioneering studies. About one hundred years ago, Ebermeyer (1876) measured the amount of leaf and branch litter in forests of important tree species in Germany, determined their inorganic composition, and analysed the effect of litter removal (which was then a widespread practice), on the properties of forest soils and growth of forest trees. He published the results of his work in a volume on what we call now "mineral cycling in forest ecosystems". According to Adams (1928), who studied the relationships between the amount of leaf and production of wood in pine forests, R. Hartig discussed the same subject in 1891. Boysen Jensen (1910) analysed shade tolerance of forest trees in the context of the balance of production and consumption of organic matter. Boysen Jensen (1927, 1930) also studied primary production, including gross production, of young stands of ash and beech and discussed the problems of thinning of forest stands. Outlines of these works were included in his book "Die Stoffproduction der Pflanzen", which is one of the classics in plant science. However, his work did not attract the attention of many scientists as it was hardly referred to in textbooks of ecology and silviculture before World War II. This type of study was made on a larger scale by his compatriot Möller; and became well-known when it was cited in detail in Baker's textbook (1950). Subsequently, the interest of many scientists was attracted to this kind of investigation. Moller supplemented and improved his work later with the aid of collaborators (Möller et al. 1954a, b). In Switzerland, Burger (1929-1953) worked on leaf mass and its relationship to bole wood production in forests of various species and published a series of thirteen papers which have been reviewed by Satoo (1955). In the early 1950s studies of forest productivity were initiated in several parts of the world. In Japan the biomass and wood production of pine plantations of different spacing were studied by the University of Tokyo (Satoo 1952, Senda et al. 1952). Later similar projects were started at the Government Forest Experiment Station at Meguro (Sakaguchi et al. 1955). In 1958, the 'joint research project of four universities' was started. In this project scientists from the Universities of Hokkaido, Tokyo, Kyoto and Osaka City worked together in the same forests. The data collected were considered as the common property of the scientists in the project and were distributed to interested people at some other universities because the field work was a time-consuming job and it was believed that there were many ways to analyse the same data. After the beginning of the International Biological Program, the number of research workers increased, and large amounts of data were accumulated, mostly by the harvest method which requires little apparatus. Work on these lines was reviewed and synthesised by Kira and Shidei (1967), Tadaki and Hatiya (1968), Satoo (1968b, 1970b, 1971b), and Yoda (1971).

In Britain, Rennie (1955) synthesised early data on forest dry matter production and nutrient uptake in an effort to predict the effects of planting forests on poor moorland sites. About the same time, Ovington (1956) collected new data from a number of species trials to ascertain the nutrient demands of afforestation with exotic tree species. This led to more detailed studies of pine plantations and natural birch stands (Ovington, 1957; Ovington and Madgwick, 1959a, b). Ovington (1962, 1965) has reviewed and synthesised much of the earlier work.

In Russia, the study of dry matter production of forests was also coupled with estimates of nutrient uptake (Remezov, 1959). Russian interest in the geographical variation in forest ecosystems led to a synthesis of worldwide literature by Rodin and Bazilvich (1965) whose book was later translated into English (1967).

Scientists from the main centres of forest biomass research frequently extended their interest to foreign countries, especially to the tropics, where a number of early studies were undertaken by, for instance, Ogawa et al. (1961), Greenland and Kowal (1960) and Rozanov and Rozanova (1964 quoted in Rodin and Bazilevich).

2. FORESTS

A forest may be defined as a collection of trees occupying a certain ground area and forming an ecosystem together with many other living and dead organisms in an inorganic environment including the mineral soil and atmosphere. The components of such an ecosystem interact with one another. To be called a forest, it is indispensable that the trees influence each other and are not independent as in a park or savanna. The crowns of the trees in the ecosystem make a layer called the crown canopy, and when they make a continuous canopy we say that the forest is closed. Trees growing in a closed forest and making a crown canopy have a very different shape from isolated independent trees (Honer 1971). When grown in isolation, foliage is produced almost to ground level, and, as such trees mature, this foliage increases exponentially (Madgwick et al. 1977). However, when grown in a closed forest, tree crowns are shaded by neighbouring trees and the amount of foliage for a tree of given diameter is greatly reduced compared with open-grown trees of the same species.

The amount of foliage over a unit of ground area in a closed forest is determined by the degree of mutual shading, the shade tolerance of the tree and the longevity of the foliage. While passing through the canopy, solar radiation is absorbed and reflected by leaves, and, to a lesser extent, the branches and boles. Radiation intensity at a given level in the crown canopy (I) depends on the leaf area per unit ground area above (F). This relationship has been described by an equation similar to the one for Lambert-Beer's Law (Monsi and Saeki, 1953), as

$$\log (I/I_o) = -KF \dots\dots\dots\dots\dots\dots\dots\dots\dots (2.1)$$

in which I_o is the radiation intensity above the canopy and K is the extinction coefficient which differs among species depending on the nature and angle of orientation of their foliage. This relationship which was originally found in communities of grasses and herbs, holds true also for forest canopies. Leaf weight can be used instead of leaf area (Satoo, 1962b). Figure 1 shows this relationship for young pine plantations of different stand density but the same age. In the forest, branches have a strong effect on the extinction of light and the value of K is increased with decreasing stand density since less dense stands have more branch mass per unit area of forest. Leaves in the lower parts of the canopy do not receive enough radiation to maintain high rates of photosynthesis. Table 1 includes apparent photosynthesis during the daytime for upper crown leaves exposed to direct sunlight (sun-leaves) and lower, shaded leaves (shade-leaves). Leaves in the lower, shaded crown always show lower photosynthetic rates than those in the upper crown. Sometimes shade leaves have negative photosynthetic rates which means that consumption of organic matter by respiration is larger than production by photosynthesis. As leaves consume organic matter by respiration in the nighttime and release carbon dioxide, the figures in Table 1 overestimate the daily balance of organic matter. If net photosynthesis remains negative over a period of time leaves cannot produce enough material for their own support and die. Thus, if the amount of leaves in the canopy exceeds a certain limit, the leaves in the lowermost canopy cannot receive enough radiation to replace the photosynthate consumed by themselves, by their supporting branches, and in the production of new leaves. As a result the branch dies. From equation 2.1

$$F = -(1/K)\log(I/I_o) \dotfill (2.2)$$

If the minimum intensity of the radiation to keep branches and leaves surviving and reproducing themselves is I_{min}, the upper limit of the leaf mass in the crown will be F_{max}, and

$$F_{max} = -(1/K)\log(I_{min}/I_o) \dotfill (2.3)$$

It is reasonable to assume that for any site, stocking and tree species I_{min} and K will be constant and that I_o is more or less constant, on average, for a given locality. It follows that F_{max} will also be relatively constant for a given tree species, stocking and site. Since there is an upper limit to leaf mass, trees growing in closed forests lose the lower part of their crowns through death as new growth is added to the upper crown. The canopy as a whole moves upwards. The amount of foliage in the canopy varies greatly among tree species (see

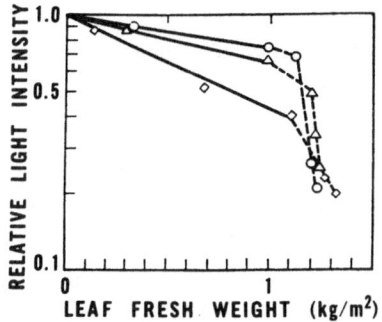

FIGURE 1. The relationship between relative light intensity at various depths within the crown canopy and to the amount of foliage above for young plantations of <u>Pinus densiflora</u> of different density (Satoo 1968). Trees/hectare and K values are o 7441, 0.2885; △ 4011, 0.4184 and ◇ 2462, 0.8414 respectively.

Table 1. Daily photosynthesis by "sun leaves" and "shade leaves" of <u>Cryptomeria japonica</u> in a closed stand (mg CO_2/g). Respiration during the night was not taken into account (Takahara 1954)

Sample tree no.	Date	Weather	Sun leaves	Shade leaves
1	Sept 27	Clear	4.07	0.42
2	Sept 28	Clear, occasionally cloudy	2.09	0.29
3	Sept 29	Clear, occasionally cloudy	3.12	0.19
4	Oct 1	Cloudy	10.47	-0.62
5	Oct 2	Clear, occasionally cloudy	8.57	1.65
6	Oct 3	Cloudy	5.49	-0.53

Chapter 4), as does the amount of solar radiation reaching the ground. Under the canopy of those tree species having large amounts of foliage, it is relatively dark and other green plants can hardly exist. In contrast, under the canopy of forest tree species with less foliage, many other green plants receive enough solar radiation to grow vigorously (Fig. 2). Figure 2A illustrates <u>Larix leptolepis</u>, an example of a tree species having a small amount of foliage. Under the canopy of large trees there are layers of smaller broadleaved trees,

shrubs, and ground vegetation. Figure 2B shows the canopy of a forest of Thujopsis dolabrata, a conifer with a very large foliage mass. It is so dark under the canopy that hardly any green plants are found there, and the forest consists of practically only one layer. Between these two extremes, there are many different types of forests. Some have multiple layers of understorey including shrubs and ground vegetation, while others have only ground vegetation under the overstorey canopy of trees. It is possible to devise a forestry practice of growing more shade-tolerant species such as firs or Chamaecyparis underneath the light canopy of a tree species such as Pinus densiflora or Larix leptolepis. Monsi and Saeki (1953) gave the name 'productive structure' to diagrams showing the vertical distribution of radiation intensity, the dry weight of non-photosynthetic organs and the distribution of leaves within stands. From Fig. 2 we can see that forests have very complicated and varying productive structures.

The mass of green plants growing beneath the canopy of a forest depends on the leaf mass of the canopy; thus, as leaf area index (leaf area per unit ground area) of the crown canopy increases, the leaf area index of undergrowth including secondary layers of trees and ground vegetation decreases (Fig. 3). Figure 3A includes data on Japanese forests, both coniferous and broadleaved, evergreen and deciduous. If we use only the data on deciduous broadleaved forests, the relationship becomes clearer (Fig. 3B) and the relationship can be approximated by a simple equation of the form

$$y = a - bx \quad \dots\dots\dots\dots\dots\dots\dots\dots\dots\dots\dots\dots \quad (2.4)$$

where y is leaf area index of undergrowth and x is leaf area index of the crown canopy. For the data in Fig. 3 the value of b was 1.1. Considering the small number and the scatter of the observations, we can assume that the value of b is approximately equal to unity. In this case equation 2.4 becomes

$$y = a - x \text{ or } y + x = a \quad \dots\dots\dots\dots\dots\dots\dots\dots \quad (2.5)$$

In other words, the total sum of the leaf area in forest ecosystems with deciduous tree species as the top layer is approximately constant, though there are large differences among individual stands if we compare canopy and undergrowth separately. Leaf mass of the forest canopy is complemented by the leaf mass of undergrowth.

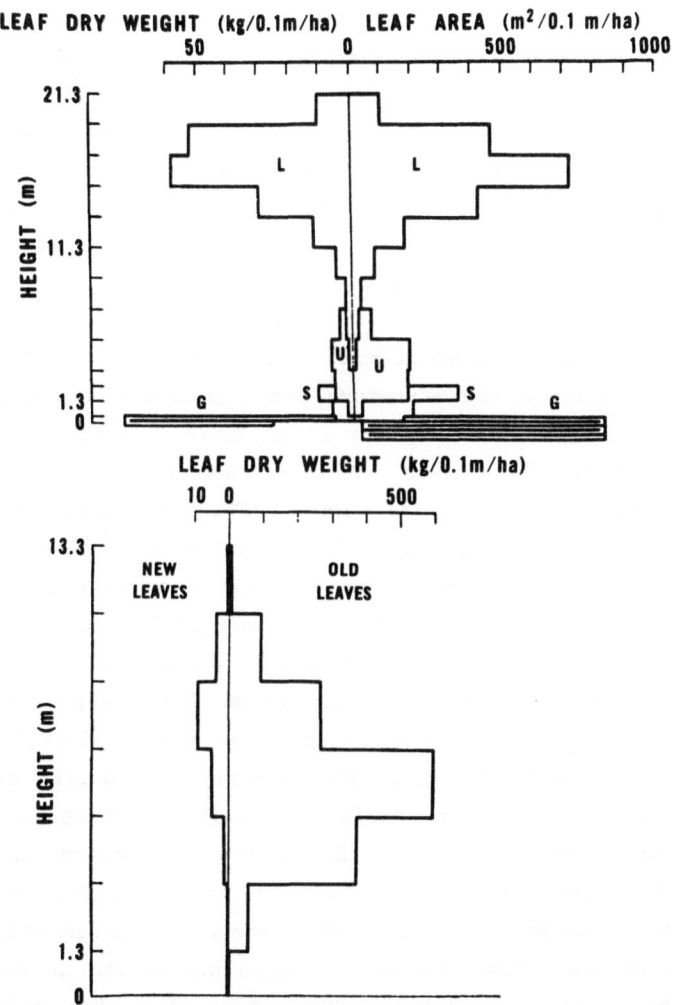

FIGURE 2. The vertical distribution of foliage in two contrasting forest ecosystems. Upper diagram: planted _Larix leptolepis_ (Satoo 1970a) Key: L _Larix_ overstorey, U understorey of deciduous broadleaved trees, S shrubs, G ground vegetation. Lower diagram: planted _Thujopsis dolabrata_ (Satoo _et al._ 1974)

The character of leaves also changes with the change of radiation intensity throughout the layers of the forest canopy. This can be seen when we study the ratio of leaf weight to leaf area within canopies. With decreasing radiation intensity, leaf area per unit weight (specific leaf area) increases (McLaughlin and Madgwick 1968). For instance, the specific leaf area in the crown canopy of a plantation of <u>Metasequoia glyptostroboides</u> increased from 12 m^2/kg at the top of the canopy to almost 30 m^2/kg at the base. This may also be true when different species are involved. For instance, in the larch stand illustrated in Fig. 1, specific leaf area was 11.8 m^2/kg for the <u>Larix</u> overstorey, 27.4 for the deciduous broadleaved second storey, 33.6 for the shrub layer, and 33.9 for the ground vegetation. Tadaki (1970) has shown that specific leaf area is related to relative light intensity but the data available to him were insufficient to distinguish between three alternative models. The change in specific leaf area parallels changes in the structure and function of leaves. The leaves in the upper layer of the canopy ('sun leaves') have lower specific leaf areas while the leaves of the lower crown ('shade leaves') have a larger specific leaf area. The structure and physiological nature of typical sun and shade leaves are described in many books on ecology and plant physiology. However, between these typical conditions there is a continuous change as seen in Fig. 4.

The environment within the forest is strongly influenced by the forest canopy. Conditions not only differ in a major way from conditions outside the stand, but spatial variation occurs within a forest both horizontally and vertically. The quantity and quality of light is different outside and within the forest owing to differential absorption of solar radiation as it passes through the crown canopy and the understorey. Air temperature is affected as both incoming and outgoing radiation are intercepted by the crown canopy. Compared with open land, temperatures are warmer during the night and cooler during the daytime, and the vertical gradient differs from that over open land. Atmospheric humidity is higher in the forest since water vapour is released from plant leaves as transpiration and the movement of air within the forest is small compared to that in the open. The

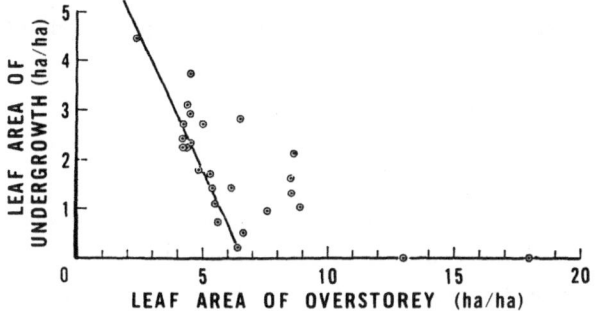

All species combined

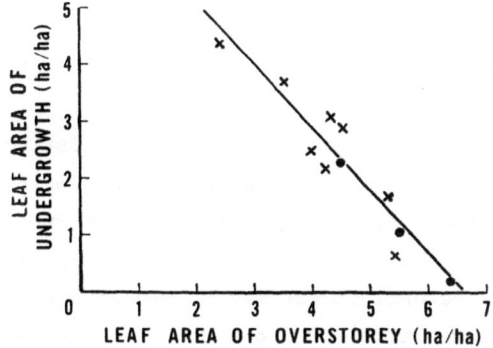

Deciduous broadleaved forests (<u>Betula</u> spp. x; <u>Fagus crenata</u> ●)

FIGURE 3. The relationship between the leaf area index of the overstorey and undergrowth (Satoo 1973).

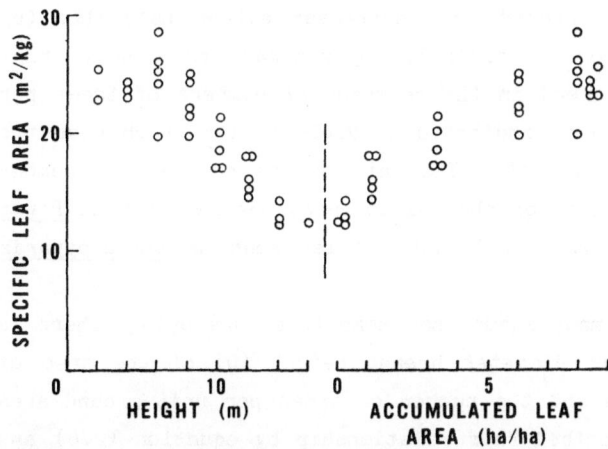

FIGURE 4. Variation of specific leaf area within the canopy of a plantation of <u>Metasequoia glyptostroboides</u>.

concentration of carbon dioxide is influenced by absorption during photosynthesis of green plants and by release during respiration of plants and animals.

The vertical profiles of the micrometeorological parameters change through the day. Thus, the maximum values of CO_2 concentration occur during the night and near ground level (Woodwell and Dykeman, 1966).

Wind velocity within the forest is low. Precipitation at ground level is also reduced by the interception of rainwater by the crown canopy. The relative reduction is greatest for light showers when the soil underneath trees and shrubs may be dry whereas in the open the surface soil may be thoroughly wet. On the other hand, when the atmosphere is supersaturated with water vapour, leaves and twigs in the canopy trap water and make so-called "wood rain", and increase moisture supply to the soil. Soil properties are also changed greatly by the existence of forests (Ovington 1953).

As most trees making the upper layer of a forest have distinct trunks, the number of individuals present may be counted. In coppice, where individual trees have many stems sprouting from a single stool, we may describe the stand by the number of stools per unit area. The number of stems per unit area of a natural forest may be enormous while the trees are small. For instance, more than one hundred thousand stems of natural regeneration have been found on a hectare of forest land (Madgwick and Kreh 1980). However, as the dimensions of individual trees increase and the requirement for space increases, the number of trees per unit ground area decreases either naturally (by so-called self-thinning) or artificially (by man-made thinning). Fig. 5 shows examples of the trend in the decrease of numbers of trees per hectare in managed forests as indicated by yield tables which give the average pattern of forest growth. The rate of the decrease of number of trees is faster in forests of shade-intolerant species such as Pinus densiflora compared with shade-tolerant species such as Chamaecyparis obtusa.

In forests having so many stems that some trees are dying, there is a relationship between the diameter breast height ($\bar{D}$) of the tree of mean cross-sectional area and the number of trees per unit ground area (N). Reineke (1933) described this relationship by equation (2.6) and named the line the full-density curve.

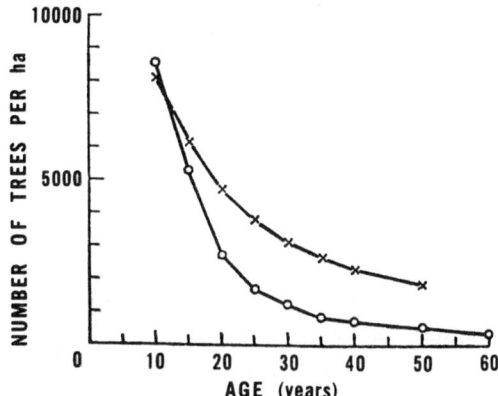

FIGURE 5. The change in number of trees per hectare with stand age. (Key: <u>Pinus densiflora</u> o; <u>Chamaecyparis obtusa</u> x)

$$\log N = -A.\log \bar{D} + K \quad\dots\dots\dots\dots\dots\dots\dots\dots \quad (2.6)$$

Reineke concluded that K is a constant differing among species and A is a constant that takes a value of 1.605 regardless of tree species. Sakaguchi (1961) determined the constants of the equation for some of the important forestry species of Japan namely

$\log N = -1.6307 \log\bar{D} + 5.5100 \quad\dots(2.7)$ for <u>Cryptomeria japonica</u>

$\log N = -1.3563 \log\bar{D} + 5.1365 \quad\dots(2.8)$ for <u>Chamaecyparis obtusa</u>

$\log N = -1.6383 \log\bar{D} + 5.3360 \quad\dots(2.9)$ for <u>Pinus densiflora</u>

$\log N = -1.7273 \log\bar{D} + 5.3773 \quad\dots(2.10)$ for <u>Larix leptolepis</u>

where N is the number of trees per hectare and $\bar{D}$ is the diameter of the tree of mean cross-sectional area in centimetres. Reineke found equation (2.6) empirically using data from many species. However, the form of equation (2.6) can be deduced from other already known relationships. Average trees from forests of a given type have similar form, and consequently it may be postulated that there is a definite relationship between diameter breast height ($\bar{D}$) and the amount of foliage ($\bar{W}$) on trees of mean cross-sectional area from stands of a comparable nature (Satoo 1962a) which takes the form

$$\log\bar{W} = b.\log \bar{D} - a \quad\dots\dots\dots\dots\dots\dots\dots\dots\dots \quad (2.11)$$

where a and b are constants. This equation is similar to the one proposed by Kittredge (1944) for the relationship between diameter breast height (D) and amount of leaves (W) for individual trees within a stand. However, the slope for the relationship among mean trees from different stands is less steep than the slope derived from trees within a single stand. Among forest stands of comparable nature, there is also a relationship between the amount of leaves on trees of mean cross-sectional area ($\bar{W}$) and the number of trees per unit area (N) of the stand (Satoo et al. 1955, Senda and Satoo 1956) namely

$$\log \bar{W} = p - q.\log N \ldots\ldots\ldots\ldots\ldots\ldots\ldots\ldots (2.12)$$

where p and q are constants. From equations (2.11) and (2.12),

$$\log N = (a + p)/q - (b/q) \log \bar{D} \ldots\ldots\ldots\ldots\ldots (2.13)$$

putting (a + p)/q = K and b/q = A, equation (2.12) becomes

$$\log N = K - A.\log \bar{D} \ldots\ldots\ldots\ldots\ldots\ldots\ldots (2.14)$$

which is identical to equation 2.6.

As the maximum amount of leaves per unit ground area (Fmax) of fully-stocked forests of a given species is approximately constant (equation 2.3) and as there is a close relationship between the amount of leaves on trees and their stem cross-sectional area, the amount of leaves on trees of mean cross-sectional area ($\bar{W}$) is approximated by

$$\bar{W} = (Fmax)/N \ldots\ldots\ldots\ldots\ldots\ldots\ldots\ldots\ldots\ldots (2.15)$$

Equation 2.15 is the same as equation 2.12 if, in the latter, q = 1 and p = logarithm Fmax. From equations (2.11) and (2.15),

$$\log N = \log Fmax + a - b.\log \bar{D} \ldots\ldots\ldots\ldots\ldots (2.16),$$

putting log Fmax + a = K and b = A, equation (2.16) becomes equation (2.6). $\bar{V}$, $\bar{D}^2\bar{H}$ or total tree weight can be used instead of $\bar{D}$ in equations (2.6) or (2.11) where $\bar{H}$ and $\bar{V}$ are the height and volume of the tree of mean cross-sectional area respectively. In these cases, equation (2.6) has the same form as the equation suggested by the 'law of 3/2 powers' of Yoda et al. (1963).

3. METHODS OF ESTIMATING FOREST BIOMASS

1. INTRODUCTION

The estimation of the weight of vegetation in the forest is a time consuming operation. Most often this work is completed in remote locations so that efficient methods need to be employed. Two distinctly separate situations can be considered. On the one hand there is the typical research situation in which samples of trees and subordinate vegetation can be weighed in the field. Alternatively, one may wish to estimate the weight of a stand using prediction equations based on sampling over a wide geographical area. Some problems are common to both situations while each approach has problems of its own. In either case it is customary to report results as tonnes per hectare. Some authors prefer grammes per centimeter square but these units are hardly conducive to a mental picture of the forest which often consists of a relatively few, large, widely spaced individual trees.

2. BIOMASS OF THE TREE LAYER.

Standard methods of estimating the volume of standing trees are described in textbooks of forest mensuration. These methods are applicable to estimates of stem weight though special problems may arise when trees smaller than those normally studied by forest mensurationists are of interest. When estimating the weights of other components of forest trees it is important to assess whether the assumptions underlying standard mensurational techniques still apply.

The steps involved in estimating stand weight are:

(i)   selection of a sample plot
(ii) measurement of a limited number of dimensions on all standing trees in the plot
(iii) in the case of research plots, selection and detailed measurement of sample trees
(iv) application of the values for sample trees to the remaining trees to obtain estimates of the weight of trees per unit area.

Table 2. Variability of stand parameters as affected by plot area in a stand of naturally regenerated <u>Pinus virginiana</u> based on 6 plots of 40 m^2 and 3 plots of 79 m^2.

Parameter	Plot size			
	40 m^2		79 m^2	
	Min	Max	Min	Max
Number of trees	20	27	42	50
Basal area m^2	0.094	0.107	0.198	0.210
Average height m	27.4	29.3	27.9	28.6
Biomass kg				
Leaves	15.9	20.6	34.1	40.5
Live branches	43.7	67.1	94.2	111.8
Dead branches	12.1	20.3	25.6	33.4
Stems	211.0	244.3	445.5	455.4

2.1. Selection of a sample plot

It is not normally possible to measure the whole of a given forest stand so, in general, sample plots are taken to represent the forest. In the context of research, such sample plots may coincide with measurement plots in a designed experiment or may be representative of particular stages in the development of a forest crop. The cost of physically weighing trees is such that the number of plots sampled is usually small and many published studies are based on 'typical' plots with uniform composition and little variation in topography. There is no firm basis for deciding on a suitable plot size. One suggestion is that the smallest side of a rectangular plot should be larger than the height of the dominant trees. The effect of plot size on variability of estimated weight is illustrated in Table 2 using data from a stand of naturally regenerated <u>Pinus virginiana</u> trees. All 136 trees were mapped and weighed on a 237 m^2 plot. All biomass components were more variable than conventional mensurational parameters of basal area and mean height which suggests that larger, or more, plots should be used for the estimation of biomass than would be acceptable for conventional mensurational techniques. Canopy component weights were more variable than stem weights which suggests that part of the problem could be the failure of the plot area as determined by the location of tree stems to reflect site occupancy or canopy spread. Young (1973) found similar results in apparently homogeneous stands and concluded that 'an inordinate number of plots would be necessary to evaluate a forest stand in terms of fresh or dry weight'. Doubling plot size materially decreased plot to plot variation. Such doubling substantially reduces plot area to plot edge ratio. Plantations would be expected to be less variable than naturally regenerated stands. Newbould (1967) suggested that sample plots in uneven-aged or mixed forests should be not less than 0.5 ha while in tropical rain forest the area might have to be five times that size. Since the coefficient of variation (CV) is related to plot area (P) by the relationship

$$(CV_2)^4 = (CV_1)^4 \cdot P_1/P_2 \quad \dots\dots\dots\dots\dots\dots\dots \quad (3.1)$$

(Freese 1961) it would be possible to select a suitable plot size by trial and error using some easily measured stand parameter such as basal area as an indicator. For height and diameter, small plots of Pinus taeda and P. elliottii were found to be more efficient than large plots, especially in young plantations (Conkle, 1963).

2.2. Measurements on all trees

Within a single plot there is usually a high correlation between stem diameter, height and weight. This suggests that the most efficient measurement procedure is to measure the diameters of all trees and the heights and weights of a subsample. Regressions of height and weight on diameter can be used to estimate the height and weight of the remaining trees. A variety of forms of regression have been used for estimating height from diameter. Garcia (1974) compared 38 different equations using data from 18 stands of Pinus radiata. No one relationship was clearly superior but overall the best performer was

$$H = a + b.\exp(-0.08D) \quad \dots\dots\dots\dots\dots\dots\dots\dots\dots \quad (3.2)$$

where H was the height (m), D was diameter breast height (cm) and a and b were constants. The estimation of stand weight will be dealt with in section 2.5).

2.3. Selection of sample trees

The number of trees which can be sampled in detail is usually strictly limited by time and manpower available. As a consequence researchers have tended to restrict sampling to trees of good form. Any bias which such sampling may have on estimates of forked or damaged trees has not been assessed.

Within a stand, tree weight is usually closely related to size, increasing proportionately to diameter raised to a power between 2 and 3. Stratified sampling based on stem diameter can reduce the number of sample trees required to predict stand weight within given limits. The effect of stratification depends on the method of calculating stand weight from sample tree data.

Trees on the edge of stands obviously differ from those in the interior (Zavitkovski 1981). The required size of buffer strips in biomass sampling appears not to have been studied.

2.4. Measurement of sample trees

Sample trees should be measured before felling in an identical way to those trees within the sample plot for which weights are to be estimated.

Whether sample trees are selected from within or outside sample plots will depend on the nature of the study and will affect the statistical treatment of data. In those studies in which the objective is to develop weight prediction equations for application over a wide geographical area, the distribution of sample trees will often be clustered so as to reduce travelling times. Again, such sampling should influence the statistical analysis of the data as will be discussed later.

The objectives of any particular study will tend to affect the detailed procedures for handling trees. There is no consensus concerning the components to be determined and only general guidelines can be given. Alemdeg (1980) has prepared a manual for data collection under Canadian conditions which could form a useful basis of standardization.

With small trees the whole tree can be subdivided into its components, dried and weighed. As tree size increases this becomes increasingly more difficult to the point where subsampling becomes inevitable. Several authors have used the individual branch as a unit for subsampling, estimating the total weight of the crown in ways analogous to those used for estimating stand weights from sample trees (Cummings 1941; Attiwill 1962, 1966). Most authors use branch diameter as the independent variable for estimating branch weight. Riedacker (1968) noted that, for shade tolerant species, the addition of relative height of branch within the crown materially improved the precision of estimating equations. With _Pinus radiata_ such methods tend to give biased estimates of total crown weights with the degree of bias depending on the form of the regression used (Madgwick and Jackson, 1974). The best estimating equation was of the form

$$\log W = a + b. \log D^2L + c.RH + d. RH^2 \quad \dots\dots\dots\dots \quad (3.3)$$

where D was branch diameter, L branch length, RH was relative height of the branch within the tree and a, b, c and d were constants. The equation underestimated foliage weights on branches at the top and base of the crown and overestimated them for mid-crown positions. Omitting relative height gave larger overall biases and reversed the sign of the bias with respect to crown position.

The following subdivision of sample trees and the methods of handling components are common. Division of the tree into at least crown, stem and roots is almost universal though many people leave the root system unmeasured. Branches, stems and roots may be subdivided depending on a diameter classification which, for the above-ground tree, may be related to utilization limits and, in the case of roots, to a limit determined by the ability of an extraction procedure to recover roots of a given size. In the case of evergreen trees some workers separate foliage by age class. The ease with which such a division can be made varies greatly from species to species but will often be found to yield biologically important information since leaf retention is often affected by the environment or experimental treatment. In some conifers independent estimates of needle retention can be obtained by counting the numbers of needles and needle scars of sample branches. Separation of foliage from twigs is often easier after initial drying though this might conceivably affect nutrient distribution between the two components. Stems may be weighed whole or after cutting into convenient lengths and discs removed to determine moisture content. In large stems, volume may be estimated using conventional mensurational techniques and short sample logs used to obtain volume to weight conversion factors.

Crowns may be subdivided into separate layers in order to describe the vertical distribution of foliage within the canopy or to estimate the growth of branches.

Major roots can be recovered by winching following preliminary excavation of soil around the stem. Experience suggests that this technique is viable for removing roots down to a diameter of about 5 mm. These methods fail to recover fine roots which are usually sampled using coring techniques. Rootlet distribution is very variable and a large volume of soil must be examined.

When subsampling is necessary one must first obtain the fresh weight of each component. Moisture content can then be obtained for the subsample. Even when the whole tree is dried it is useful to obtain fresh weight as the resultant estimates of moisture content provide a check on the field data. For instance, for Pinus radiata, we expect the ratio of dry weight to fresh weight to be about 80 per cent for dead branches and about 45 per cent for live branches. Moisture content varies systematically up the stem and we check these values by fitting a quadratic equation of height on moisture content. The temperature at which samples are dried will affect estimated moisture content. Forrest (1968), working with Pinus radiata, found that as drying temperature was increased from 70° C to 105° C branchlet samples lost a further 2 per cent moisture. This compared with a 7 to 8 per cent loss of initial dry weight when samples were stored for 45 days at either room temperature or in a cold room kept at 7° C. Barney et al. (1978) noted a 3 per cent difference between drying at 65° and 103° C for Picea mariana.

2.5. The estimation of stand weight from sample trees

A wide variety of techniques has been used to estimate stand weight from sample trees. Several studies have related the estimates using different techniques while a few studies have compared estimated weights against the measured values for whole plots (Ovington et al. 1967, Madgwick 1971, 1981; Madgwick and Satoo 1975; Barney et al. 1978). Details of the plots we have used for these comparisons are given in Table 3 which may be consulted in connnection with the following discussion.

The 'mean tree' method is based on the detailed measurement of one or more trees in the sample stand which are judged to be average in size. Most users of this method have taken trees of mean basal area. Plot values are calculated using this formula.

$$W = N. \bar{w} \qquad \dots\dots\dots\dots\dots\dots\dots\dots\dots\dots\dots\dots (3.4)$$

where W is plot weight, N is the number of trees in the plot and $\bar{w}$ is the weight of the average tree. The advantage of this method is its computational simplicity. Results from the nine test plots indicate that the mean tree method can give large underestimates of canopy weights (Table 4) though, for the two pine plots containing more than 100 trees, estimates based on the several trees closest to mean basal area gave results which were mostly within 10 per cent of actual plot values (Table 5).

Table 3. Summary of sample plot data. [+] = at base of stem; [*] = data collected by the joint study group on forest productivity of 4 universities, Japan; [**] = data collected by the joint study group on forest productivity of 5 universities, Japan; [***] = data of Ovington et al. (1968) supplied by Dr J.D. Ovington and Dr W.G. Forrest (Madgwick and Satoo 1975) (N = natural regeneration; P = planted)

Species and origin		Age (yr)	Plot area (m^2)	Stems (n)	Mean diam (cm)	Mean ht (m)
Abies sachalinensis*	N	9-30	1.5	45	1.66[+]	1.12
Abies sachalinensis*	N	17-30	2	34	2.27[+]	1.38
Betula ermanii*	N	18	24	25	4.93	7.00
Cryptomeria japonica*	P	10	37.2	16	7.97	5.24
Cryptomeria japonica*	P	43	32	14	15.18	14.85
Larix leptolepis**	P	18	100	14	11.05	9.11
Pinus densiflora	N	15	20	13	7.13	6.61
Pinus radiata***	P	8	810	100	13.28	7.91
Pinus virginiana	N	19	237	136	7.54	8.65

Table 4. Mean bias and coefficient of variation of estimated plot weight using one mean tree per plot and 100 replicates of five trees for seven other methods using stratified random sampling

Method	Stems		Branches		Leaves	
	Mean bias %	C.V. %	Mean bias %	C.V. %	Mean bias %	C.V. %
Mean tree	2.9	-	-13.0	-	-13.3	-
Basal area ratio	-0.7	5.4	- 1.4	18.7	- 0.6	13.0
Regression on D^2	-0.5	5.4	- 1.4	17.9	- 0.5	12.6
Log-log regression						
Theta$_1$	1.4	6.7	- 1.6	17.6	0.5	13.9
Theta$_2$	2.9	6.8	1.9	17.5	7.1	15.0
Theta$_3$	2.7	6.7	1.5	17.4	6.0	14.4
Theta$_4$	2.2	6.6	0.2	17.3	3.5	13.8
Theta$_{5,6}$	0.8	6.6	- 3.1	17.5	- 2.0	14.3

Table 5. Plot weight using the 'mean tree' method as a ratio of actual plot weight

Species	No. of samples	Stem	Branches	Foliage	Roots
Pinus radiata	3	1.07	1.00	1.10	1.03
Pinus virginiana	4	1.03	0.96	0.86	-

When no tree of mean basal area is available then plot weight can be estimated by using the stems closest to the mean and the formula

$$W = \frac{\Sigma w.G}{g} \quad \dots\dots\dots\dots\dots\dots\dots\dots\dots\dots\dots\dots \quad (3.5)$$

where Σw is the sum of the weights of sample trees, g the sum of their basal areas and G the basal area of the plot. As a further modification of this method we can stratify trees in the plot by diameter and take trees of mean basal area of each stratum (Ovington and Madgwick 1959b).

The 'basal area ratio method' is a natural extension of formula 3.5. Several trees are sampled from the plot using either a random or stratified random selection technique. Formula 3.5 is then used to estimate plot weight. Results of simulated sampling of the nine test stands suggested that stratification provides little improvement over strictly random sampling using this procedure (Madgwick 1981). On average the basal area ratio method using five sample trees per replicate and 100 replicates per plot gave mean estimates of plot weight ranging from 98.6 to 99.4 per cent of the measured values (Table 4). The range of estimates based on replicated sampling was reasonably distributed about the mean. The basal area ratio method is computationally simple and lends itself to sequential sampling. This method is equivalent to weighted regression with weights proportional to the inverse of basal area.

Since tree weight and linear size are closely correlated, especially within individual plots, it is not surprising that regression techniques are the most widely used methods for estimating plot weight for research purposes and almost the only method used for estimating stand weights where no direct sampling has been undertaken.

Tree weight can be expressed as a function of tree size and form. Measures of size used include diameter at some point on the stem (e.g. base, breast height or just below the live crown), total height, crown length, or sapwood basal area. Tree form does not appear to have been considered. Most measures of size of trees from within plots are closely correlated. Since the development of estimating equations for management purposes involves special problems we will consider first the problem of estimating plot weight from samples taken from within or close to the plots.

The most commonly used prediction formula is that first applied to trees by Kittredge (1944), namely:

$$w = a.x^b \quad \dots \dots \dots \dots \dots \dots \dots \dots \dots \dots \dots \dots \dots \quad (3.6)$$

where w is tree weight, x is a measure of tree size and a and b are constants fitted after logarithmic transformation of formula 3.6

$$\text{Log } w = y = \log a + b.\log x \quad \dots \dots \dots \dots \dots \dots \dots \quad (3.7)$$

When estimating the weight of a plot from which sample trees have been weighed the regressor variable $\underline{x}$ is usually diameter breast height except in the case of small trees when height is a more suitable variable. In some cases diameter and height have been combined as a single variable, D^2H. Fitting equation 3.6 after logarithmic transformation has the advantage that the shape of the relationship is very flexible while the increase in the variance of weight with tree size is accounted for. One disadvantage of this method is that estimates of component weights do not usually sum to the estimates from equations relating total tree weight to size (Kozak 1970). Stratification of sampling materially improves the reliability of estimates (Madgwick 1971).

Logarithmic transformation results in equations which estimate the geometric mean weight for a tree of given size. A number of biologists have drawn attention to ways of allowing for this bias (Madgwick 1970b; Baskerville 1972; Beauchamp and Olson 1973; Mountford and Bunce 1973) quoting a variety of statistical publications. Flewelling and Pienaar (1981) present a useful, recent summary. Using n sample trees, k independent variables, and s^2, the error mean square around the linearized regression, and x as the transformed value of tree size (e.g. log diameter) they point out that the following corrections for bias (theta) are possible.

$$\text{theta}_1 = 1$$

$$\text{theta}_2 = \exp(\tfrac{1}{2}s^2)$$

$$\text{theta}_3 = g_m(\tfrac{1}{2}s^2) \text{ where g is a tabulated function, and}$$

$$m = n - k - 1$$

$$\text{theta}_4 = g_m\left[\frac{m + 1}{2m}(1 - \text{alpha}_x)\,s^2\right]$$

$$\text{theta}_5 = \exp\left[(1 - 3\,\text{alpha}_x).\frac{s^2}{2}\right] \quad \text{for alpha}_x > \frac{1}{3}$$

$$\text{theta}_6 = \exp\left[(1 - 3\,\text{alpha}_x).\frac{s^2}{2}.\frac{m}{(m + 2)}\right] \quad \text{for alpha}_x \leqslant \frac{1}{3}$$

for the simple (and usual) case where a single regressor variable is
used (i.e. k = 1) then

$$\text{alpha}_x = \frac{1}{n} + \frac{(x_i - \bar{x})^2}{\text{sum }(x - \bar{x})^2}$$

where sum $(x - \bar{x})^2$ is the sum over all the sample values of x used to
calculate the regression and $\bar{x}$ is the sample mean.

Theta$_1$ gives the uncorrected estimates found in the older
literature and, presumably, in the recent literature where no specific
'correction' is mentioned. Theta$_2$ was suggested by Meyer (1941) and
more recently by Baskerville (1972) and Mountford and Bunce (1973).
Theta$_3$ is a more precise correction factor based on the work of
Finney (1941). Using theta$_3$ gives slightly lower estimates than
using theta$_2$. In the example given by Flewelling and Pienaar (1981)
this reduction amounted to only 0.2 percent. Simulated sampling
(Madgwick and Satoo, 1975) indicated that this method overestimated
weights of stems, branches and leaves by an average of three per cent
(Table 4). Individual overestimates tended to be in more serious error
than underestimates. The remaining correction factors do not appear to
have been used in the biomass literature. Flewelling and Pienaar
(1981) present a graph of the function of g_m used for theta$_3$.
Tabulated values are in Bradu and Mundlak (1970). Both sources give
values outside the range suitable for many biomass studies. Values can
be calculated by summing the series (Finney 1941) whose kth term is:

$$\frac{m^k (m + 2k)}{m(m + 2) \ldots (m + 2k)} \left[\frac{m}{m + 1}\right]^k \frac{1}{k!} \left[\frac{s^2}{2}\right]^k$$

where k ranges from 0 to infinity. For biomass studies, where s^2 is usually small, the series converges rapidly.

Flewelling and Pienaar (1981) offer the following suggestions on the selection of a relevant correction factor. When a simple estimator is required for widespread use then $theta_1$, $theta_2$ or $theta_3$ would be appropriate. When the estimators can be incorporated in a computer program, simplicity is less important and the more complex forms of $theta_4$, $theta_5$ and $theta_6$ can be used. For use in connection with experiments the unbiased estimator, $theta_4$, would be preferred where treatment affects the regression equation. If one equation can be applied to all treatments Flewelling and Pienaar recommend $theta_5$ and $theta_6$. Replicated sampling of stands for which the weights of all trees were measured suggests that uncorrected values are as good as any (Table 4). The use of log-log regressions tended to give more variable answers than the basal area ratio method.

Confidence intervals for estimated plot weights using the logarithmic form (equation 3.6) can be estimated using a variety of methods. Meyer (1938) suggested that the variance of the estimated weight, V, can be obtained using

$$v = e^{y^2 + s^2/2} \dots\dots\dots\dots\dots\dots\dots\dots\dots\dots\dots\dots \quad (3.8)$$

where s^2 is the error mean square of the estimating regression. Finney (1941) suggested a more complex form and Mountford and Bunce (1973) suggested estimating confidence intervals using the Pearson system of curves. Using simulated sampling we have found very little difference in the reliability of confidence intervals based on either the Finney or the Mountford and Bunce procedures (Madgwick and Satoo 1975). Using random sampling both procedures most often overestimated the confidence interval. The simpler procedure is to be preferred.

A number of workers have estimated tree weight (w) from diameter (D) using the formula

$$w = a + bD^2 \dots\dots\dots\dots\dots\dots\dots\dots\dots\dots\dots\dots\dots\dots \quad (3.9)$$

or its equivalent where D^2 is replaced by basal area. An advantage of this procedure is that equations are additive so that estimates of component weights sum to the estimated total weight (Kozak, 1970).

Simulated sampling has indicated that this method of estimating stand weights leads to negligible underestimates and a variability among estimates which is similar to that found using the basal area ratio method (Table 4). Stratified sampling, based on stem diameter, improved estimates (Ovington, et al. 1967) and a sample of five trees based on stratified random sampling was about as efficient as a random sample of six or seven trees.

There has been general agreement in the literature that logarithmic regressions give higher estimates of stand weight than those obtained by multiplying the weights of trees of mean diameter by the number of stems per hectare. (Ovington and Madgwick 1959b; Baskerville 1965; Attiwill and Ovington 1968; Barney et al. 1978). Others have compared the fit of different regression models to sample data (Egunjobi 1968; Crow and Laidly 1980) with conflicting results.

In summary a comparison of the more common methods in current use indicates that there is general agreement that the 'mean tree' method tends to give biased underestimates. Among the various options using log-log regression the most consistent results are obtained using $theta_1$, that is 'uncorrected' values. This variation of the log-log procedure is the only one to compare favourably with either the basal area ratio method or unweighted regressions over all three components. However, the range of results obtained using log-log regressions is greater than that obtained using the same sample trees but calculating plot values from the basal area ratio method or the regression of weight on diameter squared. The basal area ratio method gives results which, on average, are very similar to those from regressions based on D^2 but involves simpler calculations and allows easily for sequential sampling.

2.6. Estimating tree weight when destructive sampling is not possible

There appears to be a widespread belief that the form of equation used to predict tree component weight from tree size from diameter breast height is of general applicability and that regressions developed in one locality have value beyond the original sampling site. This hypothesis may apply for stems and total trees (Curtin et al. 1980) but there is considerable evidence to suggest that it does not hold for canopy components. Kittredge (1944), in his original

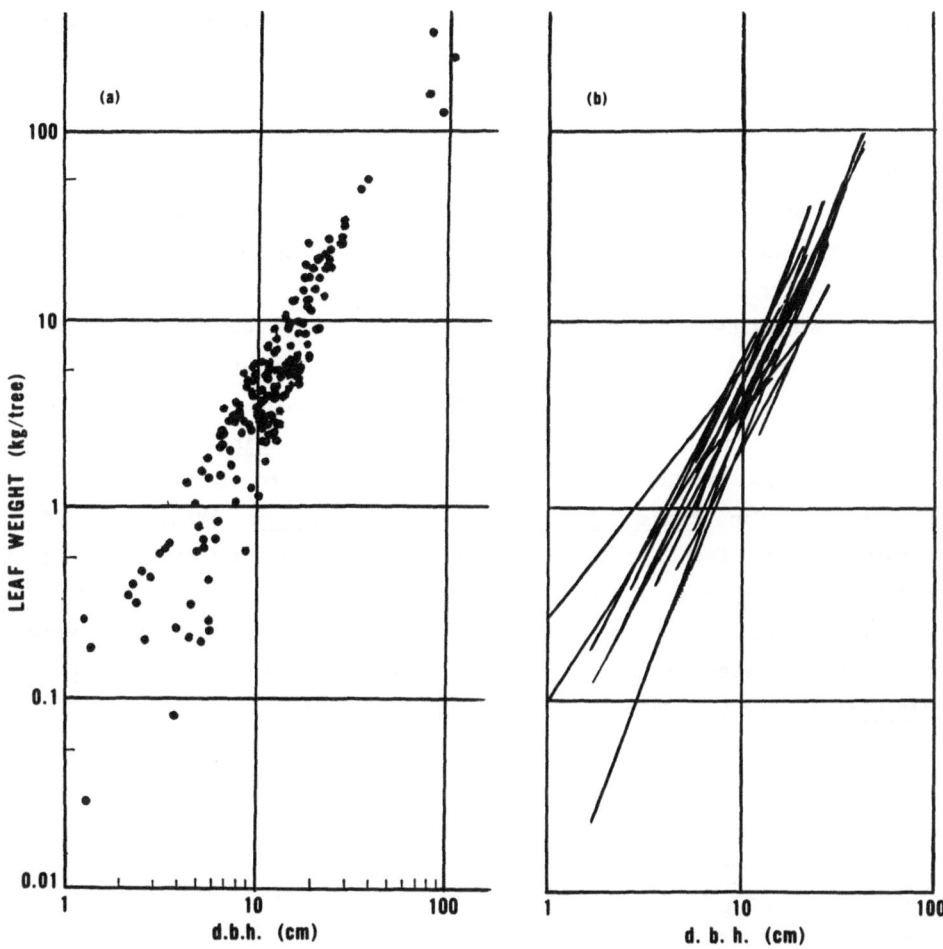

FIGURE 6. The relationship between dbh and leaf dry weight for sample trees (a) and stands (b) of Cryptomeria japonica (Satoo 1966).

publication on the subject, plotted comparative data for foliage of Pinus strobus from three sites and Pinus banksiana from two locations. In both cases trees of about 7.5 cm diameter in one location had twice the leaf weight of individuals of the same species and diameter in another location. Satoo (1962a, 1965) has compared regressions based on trees within stands and based on mean trees from different stands. He concluded that regressions based on

trees from within a stand had higher slopes than a regression based on
mean trees. (This finding is supported by Kittredge's original
study.) The intercepts of regressions based on mean trees appeared to
be influenced by management practices such as growing space. The
slopes of these regressions were affected by within-stand variation in
competitive ability, lower slopes being associated with stands of
clonal material. Tadaki (1966) has shown that the regression constants
for crown components are affected by growth stage, stand density and
site conditions.

Further evidence that diameter breast height alone is insufficient
to predict crown component weights is found in the work of Loomis
et al. (1966). They found that the addition of crown ratio
significantly improved prediction equations for Pinus echinata. We
have found that adding crown length similarly improves crown component
predictions for Pinus virginiana and P. radiata (Madgwick 1979;
Madgwick and Kreh 1980). Bunce (1968) has found significant
differences between sites using equations of log weight on log
diameter. The variability in the relationship between foliage weight
and stem diameter is illustrated for trees and stands of Cryptomeria
japonica in Fig 6. Data for clonal material is in Fig 7.

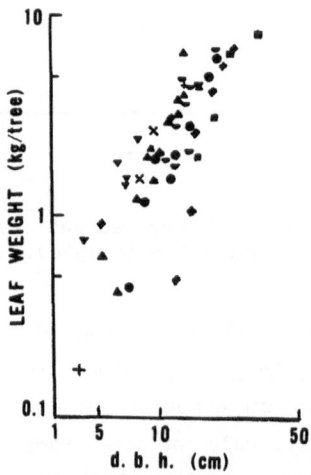

FIGURE 7. The relationship between dbh and leaf weight for
Cryptomeria japonica cultivar: Ayasugi from different stands.

Shinozaki et al. (1964) suggested that stem diameter below the live crown should be a good estimator of crown component weight since this is the point where the 'pipe' serving the crown reaches its minimum size. Storey and co-workers (1955, 1957) had previously demonstrated this empirically and their results have subsequently been confirmed by a number of other studies (e.g. Loomis et al. 1966; Madgwick 1979; Madgwick and Kreh 1980). More recently it has been suggested that sapwood area would be a comparable and equally reliable predictor (Grier and Waring 1974). It is interesting to note that Kittredge (1944), in his pioneer work, opened his paper with a discussion of foliage weight as related to stem increment before turning to the 'simpler method' of using diameter as the predictor variable even though he went on to say that 'This relationship is not as easily justified on theoretical grounds as the relation of leaf weight to periodic growth'.

Current experience suggests that a variety of forms of equation may provide a basis for generalised prediction equations for estimating tree component weights over a wide variety of sites. However, scepticism is warranted. If foliage mass is causally related to the ability of the stem to transport material we should expect factors affecting conductivity to be important as well as mere size. One might also expect the tree to have an excess transport capacity which might be related to environmental variables. It would be desirable to test these regressions on species growing over extreme climatic ranges.

Young (1976) and Hitchcock and McDonnell (1979) have summarized biomass studies containing regression equations. By far the majority were based on diameter breast height, or height or some combination of these two variables. However, other variables including crown width (Kiil, 1969), projected crown area (Rogerson, 1964), and relative crown length (Loomis et al. 1966) have also been shown to improve regressions based on diameter breast height and total height.

In a review of the statistical problems associated with the construction of biomass tables Cunia (1979a,b) noted that sample trees were usually not representative of the population, that unweighted linear regression methods were used rather than the more appropriate weighted regressions and that the underlying error structure was usually neglected. He emphasized the need to more fully explore error structures and pointed out that Schumacher (1938) forty years earlier had made the same point. Cunia (1979b) suggested two possible procedures, one using dummy variables to define clusters of samples and the other using a modified regression approach. First, calculate the regression of weight (w) on independent variables (x), namely

$$w' = b_1x_1 + b_2x_2 + \ldots \; b_mx_m \; \ldots\ldots\ldots \qquad (3.8)$$

where all the x_1 have the value 1.0. Second form new variables s and t where each s is the sum of all the weights of trees from each cluster and t_i is the sum of all the x_i. Thus t_1 becomes the number of trees in the cluster. The regression of s on t is then calculated. The difference in the error term of the two regressions is a measure of the influence of cluster sampling.

While numerous regressions have been published there appears to be a paucity of tests using an independent sample. We are currently comparing various equation forms for the prediction of stem, branch and foliage weights of Pinus radiata. Up to 247 sample trees were used to calculate regressions. An independent sample of 160 trees from 10 samplings was used to test the regressions. Logarithmic equations (equation 3.7) were used. Branch and foliage weights were best estimated from stem diameter at the base of the live crown (Table 6). Stem weights were best estimated from a second order equation in log

Table 6. Average bias of estimated weight as a percentage of measured weight based on 160 test trees of Pinus radiata using log-log regressions. For correction types see p. 24. Variable names D = dbh; H = total height; D_c = stem diameter at the base of the live crown; x = $(\log D^2H)^2$.

Dependent variable	Independent variable(s)	Correction type				
		1	2	3	4	5
Needles	D, H	39	59	59	59	58
	D_c	- 7	- 1	- 1	- 1	- 1
Branches	D, H	53	74	74	74	73
	D_c	- 8	0	1	1	1
Stem	D^2H	8	10	10	10	10
	D^2H, x	2	4	4	4	4

(D^2H). There was little difference in performance among the various correction factors $theta_2$ to $theta_6$ given above which suggests that the simplest form, $theta_2$, be used.

3. STANDING CROP OF UNDERSTOREY

Where large shrubs occur in the understorey it is possible to estimate their weights using methods analogous to those used for trees. Shrub weight can be estimated from such parameters as diameter at the root collar Brown (1976) or crown diameter. With smaller plants it is usual to harvest a number of quadrats though weight can also be estimated from ground cover (Ohmann et al. 1980). Newbould (1967) suggested quadrat sizes ranging from 100 cm^2 for mosses to 1 m^2 for uniform fine grass. Grieg-Smith (1964) illustrated the use of sequential sampling to estimate numbers of individuals of a given species and, with random sampling, the same sequential method could well be applied to biomass estimation. However, when prior information is available on the variability of quadrat data for a given site more sophisticated sampling procedures can reduce the amount of work involved. Martin et al. (1980) have shown that, in Appalachian oak forests, ranked set sampling increased relative precision by a factor of 1.66 compared with random sampling. In this method if n quadrats are to be harvested n^2 quadrats are selected and each set of n quadrats ranked visually in order of their estimated biomass. Then plot 1 is harvested in the first set, plot 2 in the second, and so on.

4. MEASUREMENT OF LITTER FALL

A knowledge of the quantity and timing of death of plant parts is important for an understanding of the functioning of ecosystems. Within deciduous forests producing one distinct flush of foliage growth a year the measurement of leaf litter can provide an economical method of estimating foliage mass. Axelsson et al. (1972) compared leaf numbers in litter fall traps with direct assessment of leaf numbers on standing shoots of Corylus avellana. The litter fall procedure was the more efficient in terms of man power.

In most species, except some young Fagus and Quercus trees, leaves fall soon after death. However, dead branches are often retained on the tree for a considerable period of time. Similarly, boles may remain standing long after death. It is theoretically possible to determine the total quantity of plant material dying in a given time period by adding litter fall and the increment of dead material accumulating on the standing trees. However, measurement of the quantity of dead material on standing trees is rather difficult and tends to be associated with large sampling errors.

Litter fall is usually measured using a number of litter traps within the stand. Big branches may be collected from areas cleared of existing litter and covered with plastic net. For twigs and leaves a wide variety of sizes and shapes of litter traps have been used in the past. The design of the trap may influence the quantity of litter trapped. For instance, square ones tend to give slightly larger values than round ones (Kirita 1967; Sasa and Satoo 1969; Saito and Shidei 1972). The number of traps necessary for any given degree of precision is dependent on crown structure, interval of collection and season. To obtain mean leaf litter fall within $\pm$ 5 percent requires from 10 to 20 traps. (Sasa & Satoo 1969, Olson 1971). Dead plant material begins to decompose before it falls from the tree (Kirita and Hotsumi 1968), with loss in dry weight of leaves ranging from 20 to 40 percent during yellowing and abscission (Bray and Gorham, 1964). Variation in total litter fall from year to year may be large. Bray and Gorham (1964) report maximum to minimum ratios ranging from 1.1 to 5.1 with an average of 2.6 for evergreen and 1.4 for deciduous species. On the other hand Olson (1971) was unable to detect any year to year variation in 11 stands studied for three consecutive years. The litter recovered may be separated into components such as leaf (by species, if necessary and possible), branch, bark, dead insects and other small animals and their frass. Inseparable material is often classified as "other". Separation of litter is both tedious and time consuming.

5. ESTIMATION OF GRAZING

The consumption of forest biomass through grazing is usually assumed to be negligible and ignored. Bray (1961, 1964) estimated foliage consumption from the percentage of leaf area consumed. He found a value of 4 to 12 per cent. Bandola-Ciolczyk (1974) found large year to year variations in foliage consumption of <u>Quercus robur</u> by <u>Tortrix</u> ranging from 8 to 39 percent. In unpublished work with <u>Fraxinus excelsior</u> 8 percent of the leaf material had apparently been consumed.

An alternative procedure is to combine estimates of frass fall with insect feeding experiments (Witkowski and Kosior 1974).

6. MEASUREMENT OF NET PRODUCTION

The estimation of the standing crop of biomass at any instant provides only one 'snapshot' in a dynamic system. To estimate net production requires a clear understanding of what is implied by the term net production - an understanding which appears to vary among individual researchers according to their academic background. Thus a physiologist would probably assume that net production was the difference between gross production and respiration (equation 1.1). On the other hand a forester may consider net production to be the change in standing crop between two measurements or, in his terms, current increment.

A practical and widely used way of obtaining net production (Pn) is by estimating the increase of dry weight of trees. Supposing that Y_1 and Y_2 are values of the standing crop at two points in time, t_1 and t_2, and that nothing is lost from the trees during the time period between t_1 and t_2, then

$$Y_2 = Y_1 + Pn \quad \dots\dots\dots\dots\dots\dots\dots\dots\dots\dots\dots \quad (3.10)$$

However, in the real world, many things are lost from Y_1 and Pn during this time period. Branches, leaves, bark and roots and, sometimes, entire trees, die and some of these materials fall as litter or are sloughed as dead roots or are grazed by insects and herbivores. So, we must clearly distinguish between the increment of the standing crop and net production. The difference between the two quantities is the amount lost by litter fall, root sloughing and grazing. The measurement of the material dying during the time interval is difficult but, for parts above-ground, is often estimated from litter fall. This assumes that the quantities of dead material on standing trees at times t_1 and t_2 are identical, or that the difference, if it exists, is accounted for. The litter generated (L) and plant material grazed (G) between time t_1 and time t_2 may have been part of either the standing crop at t_1 (i.e. Lo and Go) or from the material produced (Pn) between t_1 and t_2 (i.e. Ln and Gn) hence

$$G = Go + Gn, \text{ and } L = Lo + Ln \quad \dots\dots\dots\dots\dots\dots \quad (3.11)$$

Thus

$$Pn = Y_2 - Y_1 + L + G \quad \dots\dots\dots\dots\dots\dots\dots\dots \quad (3.12)$$

While L and G may be measured it is impossible to separate them into Ln and Lo and Gn and Go, except for material with life expectancy short compared with the time interval $t_2 - t_1$ (e.g. budscales and flowers) for which Lo may be zero. The value of G can be estimated from the fraction of leaves grazed (Bray, 1961, 1964) or from frass recovered with litter, but it is rather difficult in practice, and neglected in most studies. In most applications of the harvest method only Y_1 and Y_2 are estimated and apparent net production (Δ Y) obtained by using the formula

$$\Delta Y = Y_2 - Y_1 \quad \dots\dots\dots\dots\dots\dots\dots\dots\dots\dots \quad (3.13)$$

This underestimates real net production by (L + G). Separate equations of the form (3.12) can be used for each component of the stand (e.g. bole, Ys; branch, Yb; leaf, Yf and roots, Yr).

In that case Δ Y is estimated as:

$$\Delta Y = \Delta Ys + \Delta Yb + \Delta Yf + \Delta Yr \quad \dots\dots\dots\dots\dots\dots \quad (3.14)$$

It is possible to divide the above-ground part of trees into sufficiently small components that losses through litter fall and grazing can be at least partially accounted for. For instance, in evergreen species dividing foliage into its constituent age classes allows an estimate of loss of each age class from t_1 to t_2 as long as the time difference t_2-t_1 is of the same order of magnitude as the width of the age class. The estimation of the amount of root material dying in any time interval is particularly difficult. Estimates for such loss are high (Harris et al. 1977) and suggest that many estimates of net production are seriously biased.

Accurate estimates of increment over time for forest stands are difficult to obtain for a variety of reasons. The felling of trees at time t_1 will obviously disturb the stand except in thinning studies when trees cut as part of the stand treatment can sometimes be used to estimate stand weight. This problem can be overcome by taking trees from plot surrounds and assuming that the trees from the surround are similar to those in the plot. Errors in estimating plot weights are often large and, especially with older stands, increment can be small compared with the initial weight of the trees. A variety of methods has been used to estimate increment and these will be outlined in the following pages.

Using methods described above and equation 3.14 it is possible to estimate production when harvesting is not permitted except perhaps at the last sampling. At the time of the last measurement sample trees may be felled and the harvest method used so there is no particular difficulty in measuring Y_2. However, various difficulties arise in estimating the standing crop at previous times (Y_1). The structure of forest stands changes gradually with time so that the quantitative relationships between size and weight at time t_2 cannot be safely applied to the estimation of standing crop at time t_1. Even if the quantitative relationships at time t_1 are determined by measuring sample trees felled in auxiliary plots adjacent to the sample plots, there are sampling errors and estimation is not always successful (Satoo, 1971b).

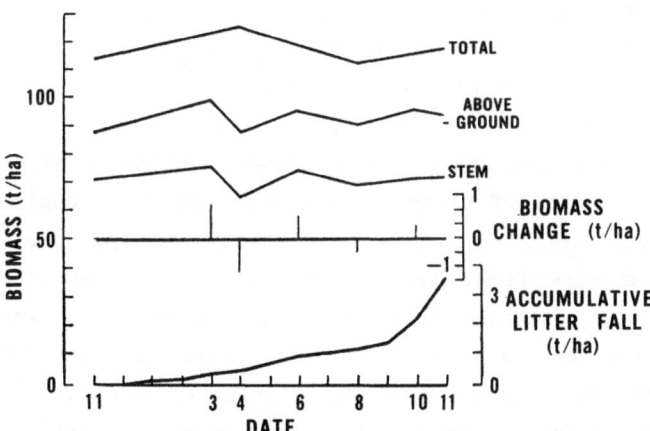

FIGURE 8. Seasonal change in the standing crop and accumulated litter fall in a young stand of <u>Pinus densiflora</u> (Satoo 1971b).

Figure 8 shows an example of the change in standing crop of a 21-year-old forest of <u>Pinus densiflora</u> estimated by non-destructive methods. Standing crop was estimated from the measurement of

diameter of all trees in the sample plot and by using equation 3.7 with
constants determined from sample trees cut from adjacent auxiliary
plots at each time. Changes in the estimated standing crop were
sometimes negative and even allowing for litter fall the negative
changes were unreasonable. An estimate of grazing was not made but
there is no reason to suppose that it was large enough to make the
changes in standing crop negative. Net production of the above-ground
parts during the time period November 1967 to November 1968 was
estimated at 7.7 t/ha whereas the comparable values using the harvest
method for approximately the same time period were about 16.0 t/ha.
The difference between the results obtained using destructive sampling
versus non-destructive methods could be caused by sampling error since,
in the non-destructive method, auxiliary plots were not exactly like
the main sample plots. Kira et al. (1967) report a successful
application of this method in a tropical rain forest.

The increment over a given time period may be measured on felled
sample trees and converted to values for a plot. Corrections for
litter fall and grazing can also be made. For woody tissues, such as
boles and branches, increment estimates would have to be made using an
extension of currently available stem analysis techniques. This
involves an estimation of stem volume at t_1 and t_2 and the
conversion of the estimated volume growth to weight growth by
multiplying by the bulk density of the new wood. The usual method used
is to sample discs from the bole several centimetres thick and every 1
or 2 m along the bole. Pairs of diameters are taken at right angles
with one measured axis being the largest diameter of the disc.
Diameter is obtained outside bark, inside bark, and, by using ring
development, at time t_1. Suitable time intervals are one year for
young planted trees but two to five years for trees having slow
diameter growth such as old planted trees and natural regeneration. A
variety of formulae can be used to convert the diameter measurements to
volume but one of the most common is:

$$Vn = \frac{1}{2} (g_n^2 + g_{n+1}^2) \dots\dots\dots\dots\dots\dots\dots\dots\dots\dots\dots \quad (3.15)$$

where g_n and g_{n+1} are the stem cross sectional areas at the base
of the nth and $(n + 1)^{th}$ bole section, 1 is the distance between
measured discs, and V_n the volume of the nth log. For the top
section of the bole the volume is calculated as a cone.

It is sometimes necessary to estimate the apex of the stem at time t_1 by assuming that ring width is uniform above the last measured disc. Wood density differs throughout the bole (Hirai 1947), so samples of wood from the increment zone in question should be measured for density (Satoo and Senda 1958). This technique is impossible for trees which do not form identifiable growth rings though marking techniques could be used to overcome this problem (Wolter 1968). We do not have a comparable method for measuring the production of bark.

6.1. Increment of branches

Measurement of the dry matter increment of branches has not often been attempted since it is a laborious job and most production studies have been made from a forestry point of view. Until recently, foresters thought that branch production was not important.

Volume increment of branches can be estimated using techniques analogous to stem analysis (Forward and Nolan 1961). The ratio of current volume increment to total volume may be used as a measure of rate of growth. Current dry matter increment is then obtained by multiplying the total branch dry weight by the fraction of volume which is in the current increment. One estimate may be made for the whole crown or the crown can be divided into separate zones.

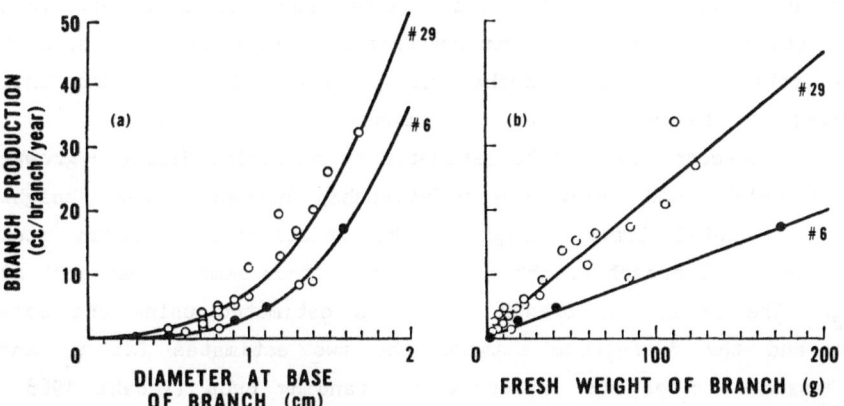

FIGURE 9. The relationship between branch growth and branch basal diameter (a) and branch weight (b) for two sample trees (numbered 6 and 29) of Cryptomeria japonica (Satoo and Senda 1966).

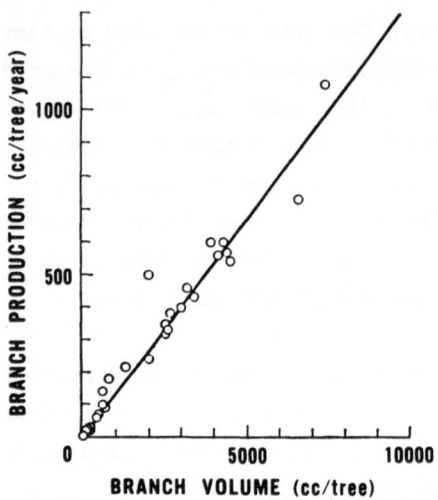

FIGURE 10. The relationship between total branch growth and total branch volume on sample trees of <u>Cryptomeria japonica</u> (Satoo and Senda 1966).

The increment of a branch within a tree is closely correlated with both branch weight and the basal area of the branch (Fig. 9). Either relationship can be used to estimate current dry weight increment of branches. However the regression constants of these relationships differ among trees and have to be determined for each individual tree within a stand. A close relationship exists between branch increment and the total amount of branches on a tree (Fig. 10) and this relationship can be used to reduce the number of sample trees.

Branch increment can also be estimated by measuring diameter growth at breast height and using the relationship between breast height diameter and total branch weight. The relationship between stem diameter and total branch weight is determined from sample trees cut at time t_2. The amount of branch at t_1 is estimated using the same equation and the difference between the two estimates (at t_1 and t_2) is assumed to be the increment of standing crop (Tadaki 1968). The amount of branch litter fall during the time period between t_1 and t_2 must be added to the estimated increment to obtain total branch production. In this method there is no evidence that the constants for the estimating equation are the same for both t_1 and t_2 and changes would be expected if competition among trees changed with time.

The relative growth rate of branch material can be assumed equal to the relative growth rate of the bole. The standing crop of branches (W_b) is multiplied by the relative growth of stem $(\Delta W_s/W_s)$ obtained from stem analysis:

$$W_b = W_b \cdot \frac{\Delta W_s}{W_s} \quad \dots\dots\dots\dots\dots\dots\dots\dots\dots\dots\dots\dots\dots \quad (3.16)$$

This method underestimates branch production (Table 7) which implies that the relative growth rate of branches is higher than the relative growth rate of bole.

Table 7. Comparison of the values of branch production (t/ha) obtained by direct measurement (A) and estimation using equation 3.16 (B)

Species	Age	A	B	B/A	Source
Cryptomeria japonica	29	2.0	0.83	0.42	Satoo and Senda 1966
Pinus densiflora	15	2.7	1.1	0.40	Satoo 1968b
Abies sachalinensis	26	4.0	1.9	0.48	Satoo 1974c
Thujopsis dolabrata	31	1.4	0.65	0.45	Satoo et al. 1974
Larix leptolepis	39	3.3	0.91	0.28	Satoo 1974b
Metasequoia glyptostroboides	17	4.3	1.5	0.35	Satoo 1974d
Populus davidiana	40	0.79	0.37	0.47	Satoo et al. 1956
Cinnamomum camphora	46	4.9	1.2	0.24	Satoo 1968a
Betula maximowicziana	47	1.04	0.43	0.41	Satoo 1974a

6.2. Foliage production

Seasonal dynamics of foliage production and loss can be very diverse. At one extreme Kawahara et al. (1981) found that on stands of Albizzia fulcata and Gmelina arborea in the Philippines there was an average turnover rate of leaves of 3.1 with leaf litter fall and biomass of foliage being about 5.2 and 1.6 tonnes/ha respectively. At the other extreme Pinus longaeva needles have been retained on trees for up to about 45 years (Ewers and Schmid 1981).

The production of foliage may be obtained by correcting the standing crop foliage for previous leaf fall of current foliage both for deciduous trees and for evergreens which lose all their old leaves just after the flushing of new leaves, like <u>Cinnamomum camphora</u> or <u>Cyclobalanopsis myrsinaefolia</u>. If the correction for premature leaf fall is neglected an underestimate will result. Litter production studies (Bray and Gorham 1964) indicate considerable variability in the pattern of premature leaf fall.

For evergreens which retain leaves two or more years the estimation of leaf production is more difficult. In species which have distinct growth flushes, such as many species of pines and firs, new and old leaves may be easily separated by age. However, in many evergreen trees, such as eucalypts it is difficult to distinguish any division between twigs produced in different growing seasons. In such cases it is not advisable to separate new and old leaves, because the numerical values tend to be accepted as "correct" and may lead to false conclusions. One solution is to mark sample twigs with paint at the beginning of the period of interest. The amount of new leaves on the tree may then be calculated by using the ratio of new to old leaves on sample branches (Morikawa 1971). The weight of current foliage may not be a valid estimate of foliage growth in evergreens since leaves have been reported to increase in weight during the second year (Kimura et al. 1968).

6.3. <u>Increment of roots</u>

We do not have good methods for estimating root production and loss though net production and net loss have both been estimated from changes in rootlet weights over time (Harris <u>et al</u>. 1977). Error estimates tend to be very large in comparison with estimated production.

6.4. <u>Increment of understorey vegetation</u>

The increment of understorey species may be estimated in the same ways as for the overstorey in the case of woody perennials. For grasses and herbs, methods developed for agricultural plants can be used.

7. GROSS PRODUCTION

Gross production of forest ecosystems has been estimated by Kimura (1960) and Nomoto (1964) using the mathematical model of Monsi & Saeki (1953). Estimates have also been based on the gradient of carbon dioxide concentration in the atmosphere over stands though one major difficulty with this method is to obtain a sufficiently large stand to satisfy assumptions in the method. Many of the estimates have been obtained by adding the consumption of dry matter by respiration to net production, as in equation 1.1. The consumption of dry matter by respiration in a forest may be calculated from the rate of respiration under standard conditions. A change of conditions affects the rate of respiration and the amount of plant tissues that respire. Negisi (1970, 1977) has reviewed the voluminous literature on this subject.

7.1. Measurement of respiration

There are many ways of measuring respiration. On a geographical basis it may be obtained from measurements of carbon dioxide concentration during periods of temperature inversion (Woodwell & Dykeman 1966). On a local scale it may be obtained from the continuous measurement of respiration of different parts of standing trees in the forest (Woodwell & Botkin 1970). The method most widely used is the measurement of release of carbon dioxide from a segment of stems, branches or from leaves in a container although the values obtained may differ from those from intact trees owing to traumatic stimulus (Negisi 1970, 1977). Carbon dioxide release can be measured either by monitoring changes in the carbon dioxide concentration in air passing through the container using infra-red gas analysis or by absorbing the carbon dioxide given off using an alkaline solution (KOH for example) in the same container as the plant sample under investigation. Subsequent titration of the alkaline solution is used to estimate the total carbon dioxide. The latter method does not need sophisticated apparatus and is easy to use in the field. Any tightly closed container could be used though an opaque container is necessary to measure respiration in plant parts (leaves and twigs) which may photosynthesize. Small material should not be packed tightly. The cut face of boles and branches should be sealed with lanoline and vaseline in the case of large samples and with paraffin for smaller samples.

Sealing is necessary to prevent gas movement through the cut surface. The effect of sealing is marked (Table 8). The volume of alkaline solution used should be proportional to the quantity of sample material. The total capacity to absorb carbon dioxide and the surface area of solution exposed are important. Three to four hours of exposure to samples is adequate and control containers including no sample plant material should be used to obtain a correction factor for the carbon dioxide content of the ambient air. The alkaline solutions are taken back to the laboratory in tightly-sealed containers and titrated to determine the quantity of carbon dioxide absorbed. The rate of respiration is corrected to a standard temperature using a Q_{10} of 2 (Table 9). To estimate respiration for a given time period, the value for the standard temperature is adjusted for the actual temperature again assuming Q_{10} = 2. However, at any one temperature, rate of respiration differs with the phase of growth within the season so necessitating measurement of respiration rate at intervals throughout the year.

Table 8. The ratio of CO_2 released from samples with cut faces sealed to samples left unsealed (Negisi 1970)

Species	Bole (Ziegler)	Bole	Branches (Prinz)	Twigs
Fagus sylvatica	12.5	-	-	-
Fraxinus excelsior	5.1	1.1	1.1	1.7
Quercus pedunculata	3.6	1.4	1.8	1.6
Acer platanoides	2.8	1.5	1.8	1.7
Betula verrucosa	1.8	1.9	1.6	0.9
Sambucus nigra	1.6	-	-	-
Alnus incana	-	2.1	3.1	1.9
Sorbus aucuparia	-	2.0	2.1	1.1
Salix caprea	-	1.8	1.8	-
Populus tremula	-	1.6	1.6	1.2
Larix europaea	1.5	1.7	1.4	1.6
Picea abies	1.4	-	-	-
Pinus sylvestris	1.1	-	-	-

Table 9. The temperature quotient (Q_{10}) for CO_2 release from
boles (Johansson, according to Negisi 1970)

Species	Q_{10}	
	5–15°C	10–20°C
Pinus sylvestris	1.82	2.10
Picea abies	2.03	2.13
Larix europaea	2.01	1.88
Larix kaempferi	2.07	2.08
Betula pubescens	2.00	2.06
Quercus robur	2.18	2.41

The total consumption of dry matter by respiration on a unit area basis is estimated oy multiplying the rate of production of carbon dioxide and the biomass for each type of plant material. It is desirable to express both biomass and respiration in the same units - either CO_2 or dry matter. Total carbon dioxide release and dry weight of plant tissue can be interconverted by making an assumption concerning the carbon content of plants. The most widely-used ratios for this are:

1 kg CO_2 = 0.614 kg dry matter .... (3.17)

assuming that starch is consumed by respiration, or

1 kg CO_2 = 0.546 kg dry matter .... (3.18)

assuming that the carbon content of plant issue is 50%. There is a difference of about 10% between the two methods, but considering the errors of the entire process of estimation this difference has been regarded as of little importance. This estimation technique involves many assumptions, particularly in the case of woody tissue. For instance, when samples are taken from different parts on the radius of boles, respiration is found to decrease from the bark towards the centre of the tree (Fig. 11). As branches and boles grow larger, the percentage of older wood with lower respiration rates increases so that respiration decreases with stem diameter (Fig. 12, 13). Even for one diameter, unit respiration varies with rate of growth, age, and environment (Negisi 1977). In order to obtain the respiration of a tree stand respiration rates for each diameter class and the standing crop of each diameter class are necessary.

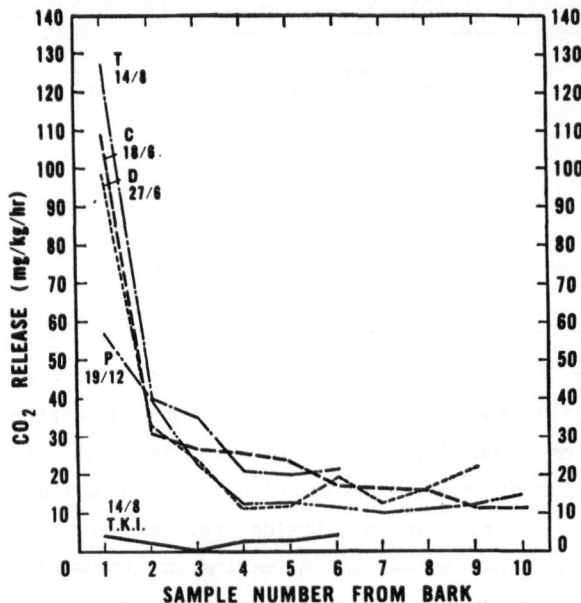

FIGURE 11. CO_2 release from different parts of the bole of <u>Fagus sylvatica</u> without heartwood (Moller 1945). Sample width 25 mm. T 14/8, etc. sample disc and sampling date. T.K.I., killed.

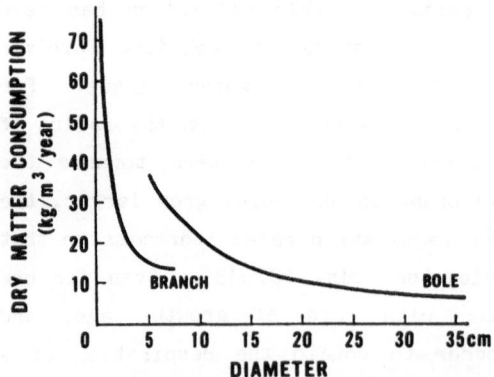

FIGURE 12. The relationship between respiration rate and diameter for branches and boles of <u>Fagus sylvatica</u> (Moller <u>et al</u>. 1954a).

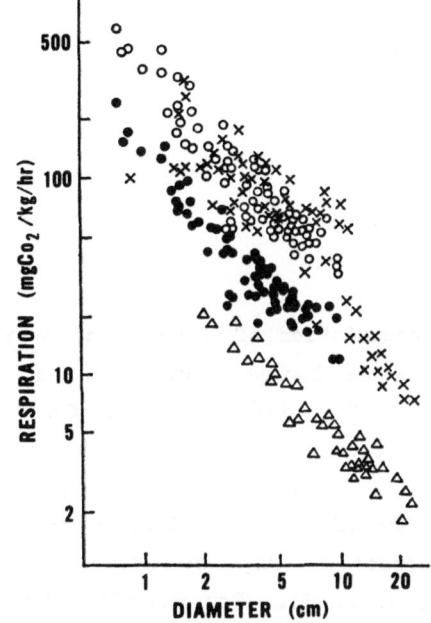

FIGURE 13. The relationship between respiration rate and stem diameter for _Pinus densiflora_ at 20°C on both a dry weight basis (x Tochiaki, May-October; ● Negisi, October) and a fresh weight basis (△ Yoda et al.1965, March; o Negisi, October).

An entirely different method for estimating the consumption of dry matter by respiration of a forest was proposed by Kira _et al_. (1960). Supposing that a forest contains a standing crop of leaves (W_L) with a mean net assimilation rate $\bar{a}$, and a total non-phosynthetic tissue (W_c) with a mean respiration rate of $\bar{r}_c$ and that these values do not change with time then:

$$\Delta W = \bar{a}\, W_1 - \bar{r}_c W_c \quad\dots\dots\dots\dots\dots\dots\dots\dots\dots\dots\dots \quad (3.19)$$

$$\bar{r}_c = \bar{a}\,\frac{W}{W_c} - \frac{\Delta W}{W_c} \quad\dots\dots\dots\dots\dots\dots\dots\dots\dots \quad (3.20)$$

where ΔW is the net production.

4. BIOMASS

Forests are characterized by the accumulation of biomass and they include the terrestrial ecosystems having the largest biomass per unit ground area. Table 10 includes examples of published data on forest stands with some of the largest reported total weights. There are many publications concerning the biomass of the overstorey tree layer, but information about the biomass of undergrowth is more limited. The biomass of the undergrowth is influenced by the overstorey both qualitatively as well as quantitatively (Fig. 3). For example, there was hardly any undergrowth beneath the canopy of a stand of Thujopsis dolabrata which carried 35.6 t/ha of leaves (Fig. 2), whereas the average weight of understorey in 20 stands of Pinus densiflora with an average canopy foliage weight of 6.9 t/ha was 40.2 t/ha. As there are often woody plants in the undergrowth which accumulate dry matter year by year, the biomass of undergrowth tends to increase with the age of the forest. For example, a 16-year-old forest of Pinus densiflora having 5.3 t/ha of leaves in the canopy had 6.2 t/ha of undergrowth biomass whereas a 55-year-old forest of the same species having 6.9 t/ha of leaves in the canopy had 40.2 t/ha of undergrowth. In managed forests, the undergrowth is often cut during management operations such as thinning, while in some forests it has been a general practice to harvest undergrowth as fuel for the local people. Consequently, comparisons of biomass of undergrowth among different forests are difficult. However, comparison of leaf biomass is easier as the lifespan of leaves is short and the effects of human interference do not last so long. Fig. 3 shows that as leaf mass of the overstorey decreases, the leaf mass of the undergrowth increases. However, there is a wide variation in understorey for any particular overstorey leaf mass depending on the type of overstorey. Biomass of undergrowth cannot be neglected since it comprised up to 27.4% of the total above-ground biomass in a range of 50 forests. Foliage biomass and foliage area of the understorey are important in determining production of organic matter in some forest ecosystems (Table 11). While the understorey may not contribute directly to timber production its role in the functioning of forest ecosystems is a neglected subject of study.

Information on root biomass is also limited, partly because of the laborious nature of determining root biomass, but also because root sampling is not always possible. Table 12 gives an example of detailed studies of biomass of each layer of a forest of Larix leptolepis. The biomass of animals in the forest may be important in terms of ecosystem function but their total mass is small especially in relation to the very large biomass of overstorey trees.

A wide variety of factors affect the standing biomass in a forest, while some of these factors may interact it is easiest to consider them individually.

Table 10. Examples of large standing crops of forests

	Tsuga heterophylla + Picea sitchensis	Pseudotsuga menziesii + Tsuga heterophylla	Abies spp. + Pseudotsuga menziesii	Evergreen gallery forest
Location	U.S.A.	U.S.A.	U.S.A.	Thailand
Approx. age years	110	100	115	–
Trees per ha	373	478	350	–
Mean height m	47.7	62.6	49.9	–
Basal area m^2/ha	98.2	63.3	98.1	–
Stem t/ha	815	601	795	467
Branches t/ha	49	49	68	214
Leaves t/ha	8	11	17	15
Total t/ha	871	661	880	696
Roots	–	–	–	54
Source	---	Fujimori et al. (1976)	---	Ogawa et al. (1961)

Table 11. Leaf biomass and area of understorey species of forests and the total amount of leaf in the ecosystem

Species	No. of samples	Total stand		Understorey	
		Dry weight t/ha	Area ha/ha	Dry weight %	Area %
Larix leptolepis	1	3.6	4.3	17.5	37.3
Metasequoia glyptostroboides	1	4.3	8.5	8.7	10.3
Pinus densiflora	20	6.9	–	9.5	–
Chamaecyparis obtusa	2	14.6	–	8.8	–
Cryptomeria japonica	2	15.1	–	7.4	–
Abies sachalinensis	1	13.8	–	(+)	–
Thujopsis dolabrata	3	35.6	14.6	(+)	(+)
Betula ermanii	4	–	3.9	–	45.6
Betula platyphylla	2	–	4.4	–	32.8
Fagus crenata	3	–	4.9	–	20.1

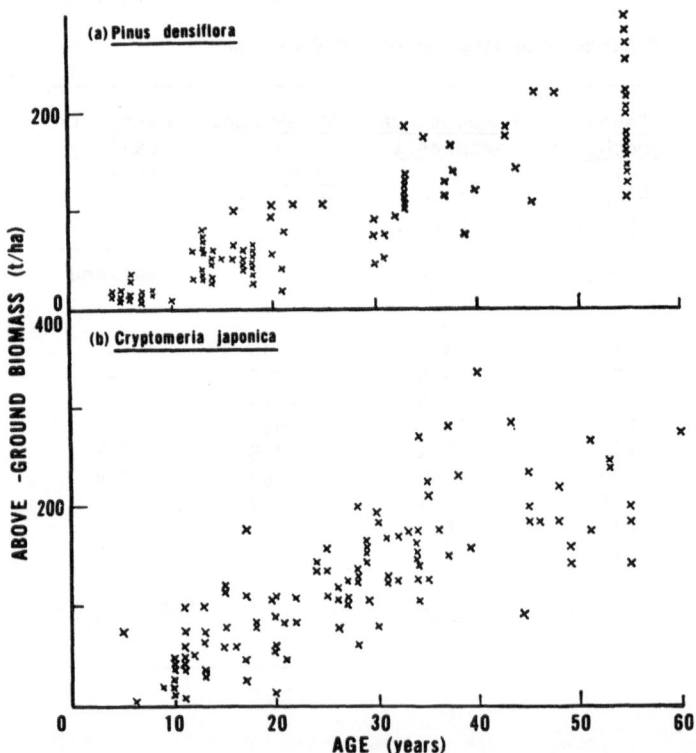

FIGURE 14. The relationship between age and above-ground biomass of overstorey forests of _Pinus densiflora_ (Satoo 1968b) and _Cryptomeria japonica_ (Satoo 1971b).

Table 12. An example of the amount and relative distribution of biomass in a forest: a 39-year-old plantation of <u>Larix leptolepis</u> (Satoo 1970a). (Figures in brackets are percentage of stand totals).

Component		Overstorey larch	Second storey deciduous broad- leaved trees	Shrubs	Ground vegetation	Total
Above- ground	t/ha (%)	164.44 (79.40)	3.20 (1.54)	0.83 (0.40)	0.96 (0.46)	169.43 (81.70)
Leaves	t/ha (%)	3.59 (1.73)	0.31 (0.15)	0.11 (0.05)	0.36 (0.17)	4.37 (2.09)
Woody parts	t/ha (%)	160.85 (77.57)	2.89 (1.39)	0.72 (0.35)	0.60 (0.29)	165.06 (79.60)
Bole	t/ha (%)	145.35 (70.11)	1.92 (0.93)	- -	- -	- -
Branches	t/ha (%)	15.50 (7.48)	0.97 (0.46)	- -	- -	- -
Underground	t/ha (%)	34.84 (16.80)	0.84 (0.41)	0.87 (0.42)	1.39 (0.67)	37.94 (18.29)
Total	t/ha (%)	199.28 (96.10)	4.04 (1.95)	1.70 (0.82)	2.35 (1.13)	207.37 (100.00)
Leaf area index	(%)	4.24 (63.47)	0.85 (12.72)	0.37 (5.54)	1.22 (18.26)	6.68 (100.00)
Specific leaf area	ha/t	1.18	2.74	3.36	3.39	-

L layer 6.7 t/ha (leaves: 5.2, branches: 1.5)
F layer 7.2 t/ha

1. OVERSTOREY TREES

The largest part of the forest biomass consists of trees forming the upper canopy. The huge biomass of forest ecosystems, as seen from Table 10, is the result of the annual accumulation of dry matter production in boles. Consequently, as far as forests are concerned, total biomass is not directly correlated with the rate of dry matter production but is primarily a function of stand age. Fig. 14 shows

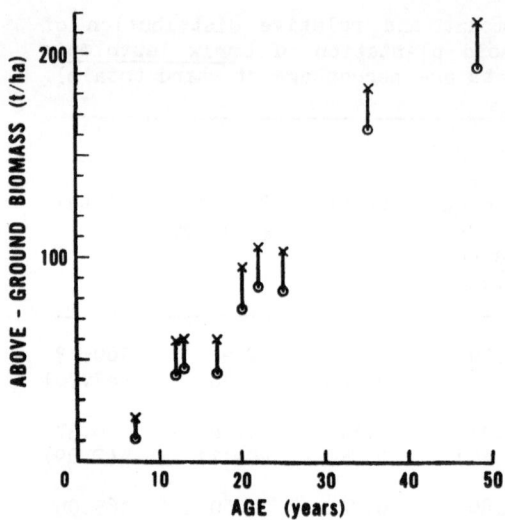

FIGURE 15. The relationship between stand age and above-ground biomass of the overstorey in a _Pinus densiflora_ forest (Hatiya and Tochiaki 1968). (x total; o boles).

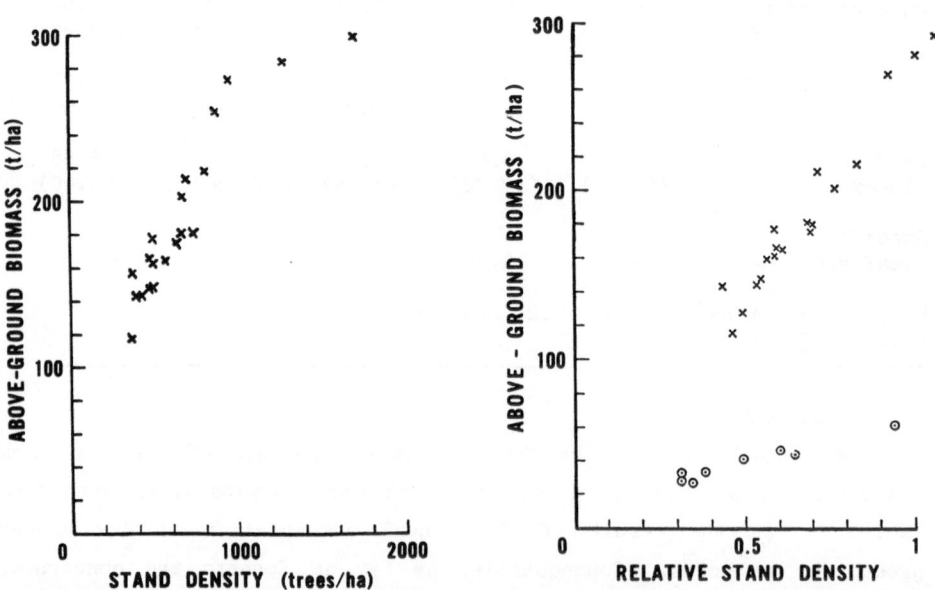

FIGURE 16. The relationship between above-ground biomass of _Pinus densiflora_ overstorey and stand density (left) and relative stand density (right). (x 55-year-old, Mori _et al._ 1969; o 14-year-old, Kato 1968).

Table 13. Biomass and estimated production of trees in a managed _Pinus sylvestris_ forest (Ovington 1957)

	Age (years)									
	3	7	11	14	17	20	23	31	35	55
Stems/ha	5310	4810	4230	5190	5640	5400	3640	2370	1890	760
Height m	0.3	1.4	2.8	3.6	4.8	5.8	8.2	12.6	14.2	16.0
Biomass t/ha										
Foliage	0.01	2.1	5.8	6.7	9.0	10.5	5.1	8.3	9.8	7.2
Branches	0.01	1.0	4.4	9.2	12.0	18.5	26.9	22.8	24.9	22.3
Cones	0	0	0	0.1	0.7	1.7	0.4	0.2	0.7	0.5
Stems	0.01	1.0	5.2	8.4	16.2	27.1	44.3	81.6	98.8	96.7
Roots	0	0.4	3.0	3.8	7.1	9.0	19.5	19.7	34.8	21.5
Rootlets	0.01	2.9	7.6	6.5	5.6	5.0	8.5	7.9	9.6	12.6
Total (A)	0.04	7.4	26.0	34.7	50.6	71.8	104.7	140.5	178.6	160.8
Estimated production (B)	-	8.1	30.9	50.0	85.4	118.9	217.7	348.4	445.2	682.9
A/B (%)	-	91	84	69	59	60	48	40	40	23

Table 14. The effect of stand density on standing biomass (t/ha) in natural stands of <u>Abies balsamea</u> averaging 9.5 m tall and 43 years after release (Baskerville 1966)

	Thousand stems per hectare					
	1.7	2.5	3.7	4.9	7.4	12.4
Foliage	17.4	16.2	17.0	18.3	18.6	19.6
Branches	18.8	17.2	17.1	16.8	16.4	16.5
Cones	0.6	0.7	0.6	0.7	0.6	0.4
Stems	70.3	68.9	78.1	92.7	105.6	117.3
Roots	29.9	29.5	36.1	38.3	41.3	46.0
Rootlets	4.3	4.6	7.4	3.9	7.8	5.9
Total	141.2	137.0	155.4	170.6	190.4	205.7

Table 15. The effect of site quality on standing biomass (t/ha) in two 40-year-old stands of <u>Pseudotsuga menziesii</u> (Keyes and Grier 1981)

	Site quality	
	Low	High
Foliage	10.0	16.0
Live branches	17.1	27.7
Stems	221.5	424.0
Roots	57.6	88.1
Total	306.2	555.8

the relationship between above-ground biomass of overstorey trees and age of managed forests of <u>Pinus densiflora</u> and <u>Cryptomeria japonica</u>. If we take a group of stands having similar growing conditions, forest biomass becomes a rather simple function of age (Fig. 15). The forests in the examples are all managed, and parts of some of the boles have been harvested and removed and branches and foliage from these trees have been left to decay. Therefore biomass is not the total production up to the point of sampling, but a net value (Table 13). Death of trees, branches and foliage and the material removed in thinnings all decrease standing biomass below the value of total production. In unmanaged forests, production and decay will tend to come into balance and when averaged over a large area the total stock of bole material will become constant.

Total standing biomass in forests tends to increase with stand density (Table 14) and site quality (Table 15). In Fig. 16 the above-ground biomass of pine overstorey of a group of 55-year-old forests of Pinus densiflora, covering a variety of soil fertility and management practices, is plotted against the number of trees per hectare. It is evident that biomass tends to increase with the number of trees per hectare. When the data are rearranged as a function of relative stand density, which is the ratio of actual number of trees to the expected maximum number of trees based on the mean diameter in the stand as described in chapter 2, the relationship is simplified and biomass becomes almost a linear function of relative stand density. The importance of relative, rather than absolute, stand density is also shown in Fig. 17, which includes data from forests of Pinus densiflora of the same age, site quality, and relative stand density but with large differences in absolute stand density. In this example the difference in biomass among stands is negligible and there is no systematic trend.

Among the dimensions of individual trees, bole diameter and volume are strongly affected by stand density, but height is normally regarded as unaffected by stand density and is used as an index of site fertility. The index of productivity or site quality, may be either the height of the stand at a given age or the height relative to the value on a standard growth curve for trees of the same age. Standard growth curves are widely available in forestry practice and show the general course of growth of forests on sites with different classes of productivity. Above-ground biomass of forests with similar age and similar relative stand density are correlated with height as in Figs. 18, 19 and 20, suggesting that above-ground biomass is closely related to site quality. Altitude (a synthesis of various environmental factors) has a large influence on forest biomass (Fig. 21).

Differences of biomass are due to differences in the rates of accumulation. Thus differences due to stand age reflect the time period of accumulation, and differences due to relative stand density

54

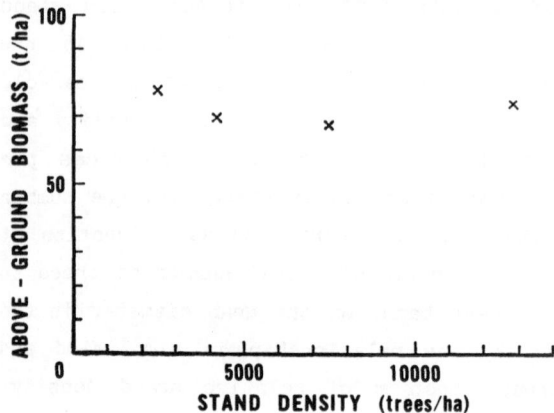

FIGURE 17. The relationship between stand density and above-ground biomass for overstorey <u>Pinus densiflora</u> stands of the same age, site quality and relative stand density (Satoo <u>et al</u>. 1955).

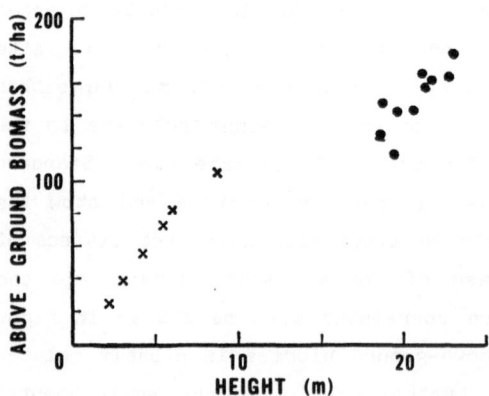

FIGURE 18. The relationship between site index expressed as tree height and above-ground biomass of <u>Pinus densiflora</u> overstorey; (x 21-year-old, Hatiya <u>et al</u>. 1966; ● 55-year-old with relative stand densities 0.43-0.60, Mori <u>et al</u>. 1969).

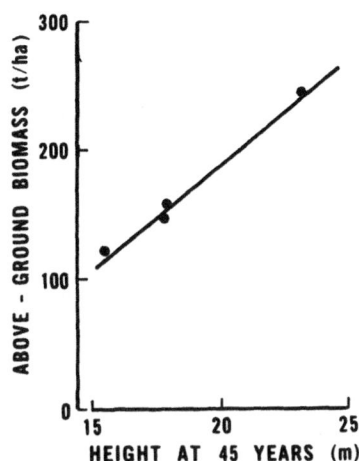

FIGURE 19. The relationship between site index (height at age 45 years) and above-ground biomass of plantation _Picea abies_ overstorey (Satoo 1971a).

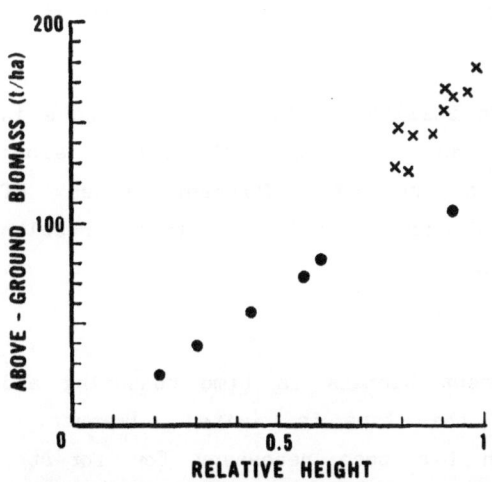

FIGURE 20. The relationship between site index expressed as relative height and above-ground biomass of _Pinus densiflora_ overstorey. (● 21-year-old, Hatiya _et al_. 1966; x 55-year-old stands of relative stand density 0.43-0.60, Mori _et al_. 1969).

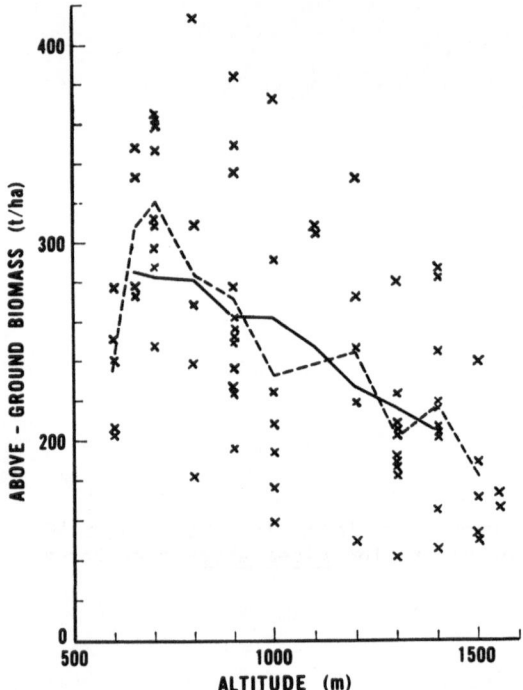

FIGURE 21. The relationship between altitude and above-ground biomass of Fagus crenata overstorey in natural stands (Maruyama 1971, & Kakubari et al. 1970).

are related to the degree of site utilisation and the differences in the removal of past production as thinnings. Differences along altitudinal (climatic) gradients reflect different rates of production. Thus, differences in biomass among forests are the results of a combination of different causes.

2. BOLE

The measurement of total forest biomass is time consuming and laborious so that information of this type is limited. However, a knowledge of stem volume growth has been necessary for forestry practice. Volume data are often published in the form of yield tables. If there is a close correlation between above-ground or total biomass of forest trees and biomass of boles, it would be possible to

approximate the biomass of a forest by multiplying total bole volume
in the stand by a conversion factor. Relationships between bole
volume and above-ground biomass have been determined in forests of
Fagus crenata, Cryptomeria japonica, and Pinus densiflora for which
a wealth of biomass data exists. For Fagus crenata above-ground
biomass can be estimated by multiplying total bole volume by 1.3
(Fig. 22). For Cryptomeria japonica and Pinus densiflora the
relationship is not so simple. The ratio of above-ground biomass to
bole biomass decreases curvilinearly with increasing bole biomass
reaching a constant value at a biomass of boles of about 80 to 100
t/ha (Fig. 23). It is theoretically possible to convert bole volume
as given in yield tables into dry weight by the use of a mean wood
density for the species available in the literature. Thus bole
weight could be used to estimate above-ground biomass. However, the
use of this kind of approximated value would be associated with a
variety of errors.

The estimation of the quantity of stem material has been an
important objective of traditional forestry. As a result there is a
large accumulation of information on factors affecting the amount of
stemwood (Assmann 1961). From Table 10 it is clear that stems may
comprise a large percentage of the total standing biomass in a
forest. The amount of stem material is affected by such factors as
stand age (Table 13), density (Table 14) and site quality (Table
15). The relationships found between stem weight and the various
environmental and management factors will closely correspond to
those found for volume in traditional forest studies though it must
be emphasized that foresters often fail to measure the total volume
of stems but measure only the merchantable part. Since cultural
practices affect the fraction of stem which is merchantable they may
affect volumes as measured by some foresters but not the total
volume and (by implication) the biomass of the whole stem.

3. BRANCHES

New branches are formed annually and parts of old branches die.
Some of the dead branch material falls to the ground but some
remains on the tree. The pattern of branch fall differs among tree
species. The amount of live branch material is affected by many

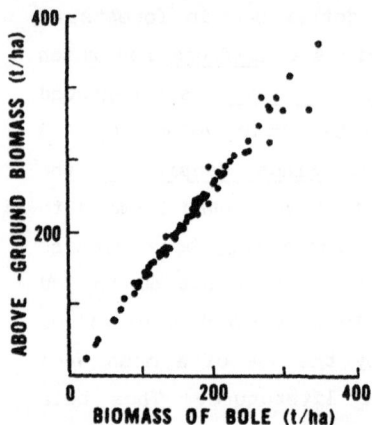

FIGURE 22. The relationship between bole biomass and tota above-ground biomass of <u>Fagus crenata</u> overstorey (Satoo 1970b).

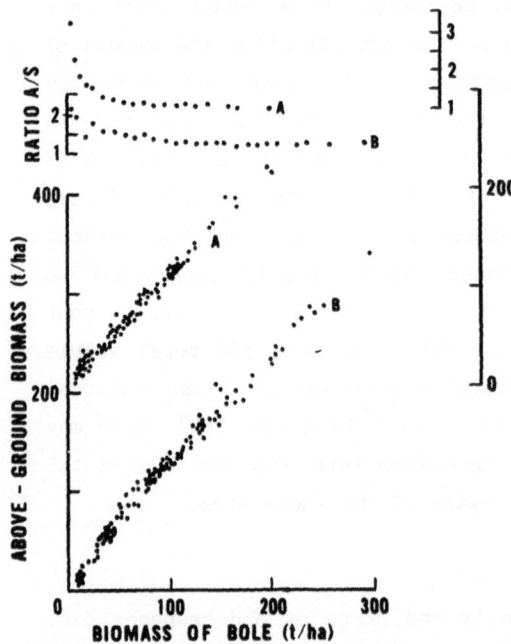

FIGURE 23. The relationship between bole biomass and tota above-ground biomass of overstorey, A <u>Pinus densiflora</u> (right har scales) and B <u>Cryptomeria japonica</u> (left hand scales) (Satoo 1971b).

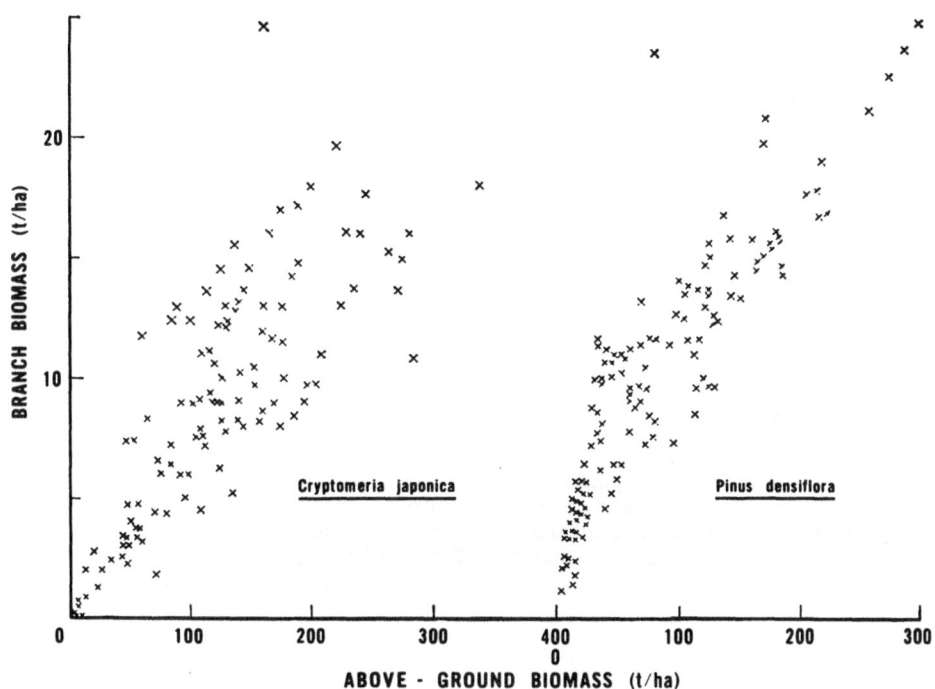

FIGURE 24. The relationship between total above-ground biomass and branch biomass for overstorey trees in forests of <u>Cryptomeria japonica</u> and <u>Pinus densiflora</u>.

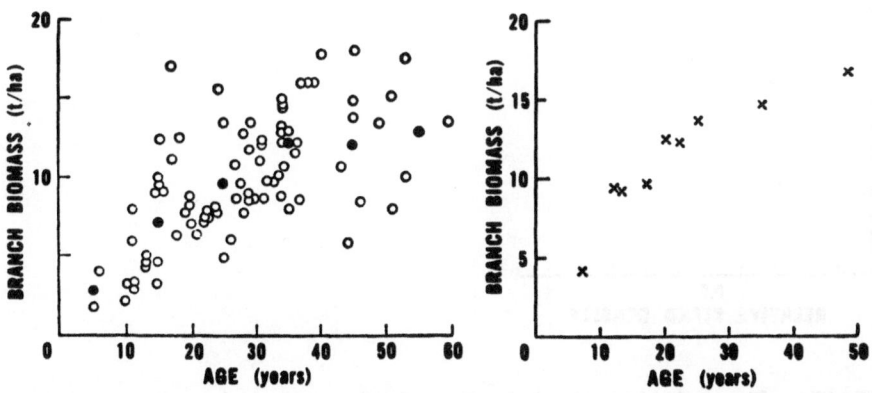

FIGURE 25. The relationship between age and branch biomass of overstorey trees in forests of, left, <u>Cryptomeria japonica</u> (Satoo 1971) (o individual stands; ● mean values for 10-year age classes) and, right, <u>Pinus densiflora</u> (Hatiya and Tochiaki 1968).

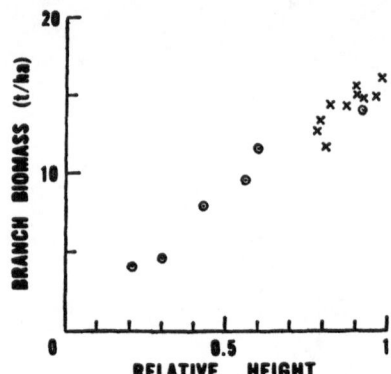

FIGURE 26. The relationship between site index expressed as relative height and branch biomass for overstorey trees in forests of Pinus densiflora (o 21-year-old stands of relative density 0.85-1.19, Hatiya et al. 1966; x 55-year-old stands of relative density 0.43-10.60, Mori et al. 1969).

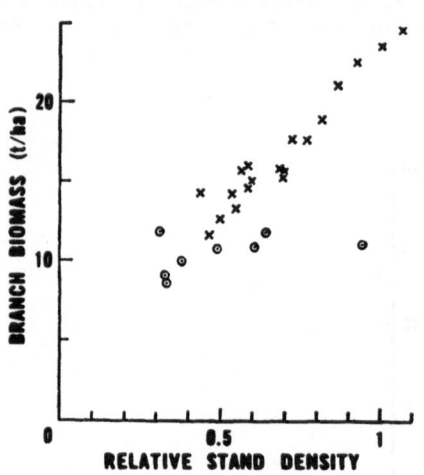

FIGURE 27. The relationship between relative stand density and branch biomass of forests of Pinus densiflora (o 14-year-old, Kato 1968; x 55-year-old, Mori et al. 1969).

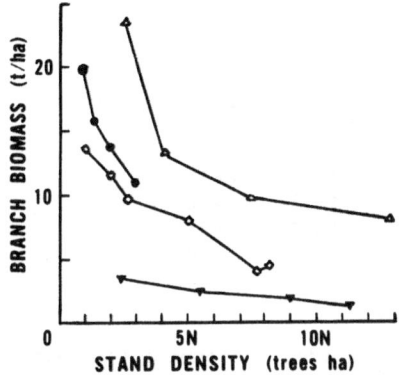

FIGURE 28. The relationship between stand density and branch biomass of forests of <u>Pinus densiflora</u> (Satoo 1968b) (▼ 4-year-old, N = 10^5; ▵ 13-year-old N - 10^3; ◇ 20-year-old N = 10^4; ● 37-year-old N = 10^3).

factors and varies widely among forest stands. Existing data of branch biomass for forests of <u>Cryptomeria japonica</u> and <u>Pinus densiflora</u> are plotted against above-ground biomass in Fig. 24. Branch biomass increases with increasing above-ground biomass, but for any given above-ground biomass, the variation of branch biomass is very large. Branch biomass increases with stand age in the early years of development but reaches a steady state in middle age (Table 13 and Fig. 25). Branch biomass increases with site quality among stands of the same age (Table 15 and Fig. 26).

The effects of stand density on branch biomass are complex and reflect stand history as well as current density. Thus Mori <u>et al.</u> (1969) found that branch biomass increased with relative density (Fig. 27) while Kato (1968) found no such effect. Data from <u>Abies balsamea</u> (Table 14), <u>Pinus densiflora</u> (Fig. 28), <u>Pinus strobus</u> (Senda and Satoo 1956) and <u>Pinus radiata</u> (Madgwick <u>et al.</u> 1977) all indicate a decrease in branch biomass with increased stocking.

Table 16. Branch biomass in the canopy of plantations of
<u>Cryptomeria japonica</u> in different regions of Japan. (Age over 30
years and relative stand density over 0.4) (Satoo 1971b).

Region	No. of plots	Branch biomass t/ha	
		Mean	Range
Tohoku, N.E. Honsyu	7	12.4	8-18
Kanto, near Tokyo	6	12.6	9-18
Kinki, near Kyoto	6	12.1	9-15
south west Kyusyu	16	12.5	6-18

In forests with high relative stand densities competition among
individual trees is severe and lower branches die earlier than in less
dense stands. Total branch biomass will build up to an equilibrium
value where production is balanced by loss through death and shedding.
This equilibrium value will be dependent on stocking. However, if the
stand is thinned or pruned the equilibrium will be upset and in stands
where tending is frequent branch biomass will never attain the value to
be found in untended stands of the same stocking. Since death and
shedding occur among the heavier, lower branches in the crown and are
difficult to predict, estimates of stand branch biomass tend to be
subject to large errors and the accurate assessment of treatment
effects on stand branch biomass is difficult to determine. For
instance, <u>Cryptomeria</u> forests in Japan are subject to considerable
variations in management practice between regions but these differences
do not appear to affect branch biomass in stands over 30 years old
which have a relative stand density above 0.4 (Table 16).

4. FOLIAGE

For many years the conclusions of Möller (1945) dominated the
thinking concerning foliar biomass. He concluded that leaf mass was
independent of various factors including site index, age and stocking
(Fig. 29). As more intensive research has been reported it now appears
that Möller's results should rather have been expressed in the form
that the two-fold range of leaf mass observed could not be simply
explained in terms of either site index, or age or stocking. As we

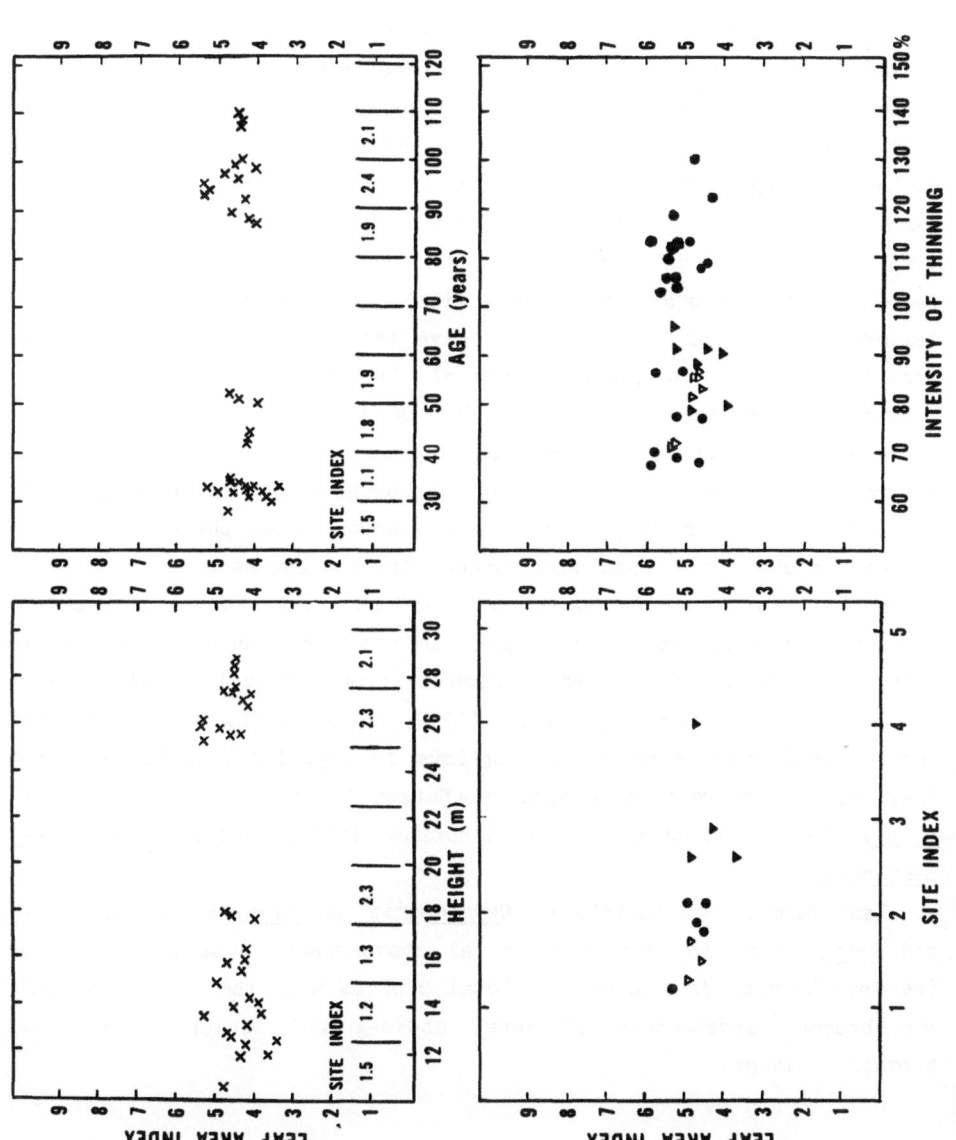

FIGURE 29. The relationships between leaf biomass in Fagus sylvatica forests and stand height, age, site index and intensity of thinning (Moller 1945).

will see in the following discussion all of these factors have now been shown to influence leaf mass but the ecological survey data of Möller has not proved susceptible to analysis in such a way as to yield information on these relationships.

When considering the effects of enviromental variables and stand treatment on foliar biomass it is important to note the methods used to derive data. In many studies separate samples of trees are taken from the various sample plots and independent estimates made on plot biomass. In other studies general prediction equations, sometimes selected from published literature and not necessarily relating to the study sites in question, are used to estimate plot biomass. In the latter approach difficulty can arise if the prediction equations are based on simple regressions of foliar mass on, say, d.b.h. It is almost inevitable that such methods will result in the conclusion that factors affecting stand basal area will also be shown to affect stand foliar biomass. For this reason I have tried to include results only from those studies using the first of these two alternatives, namely separate samples and independent estimates for each plot.

Monsi and Saeki (1953) pointed out that as leaf mass in a plant community increases, consumption of photosynthate by leaf respiration increases in proportion to leaf mass whereas total photosynthesis by leaves reaches an asymptotic upper limit dependent on radiation intensity, thus there is an optimum leaf mass for maximum production. The possible existence of an upper limit of leaf mass for the crown canopy of forests of a given species or group of ecologically similar species was discussed by Satoo (1952, 1955, 1966). Studies with agricultural crops suggest that optimum foliage levels exist and that these optima are related to species (Watson 1952). Recently, Waring et al. (1981) have demonstrated an optimum leaf amount for Pseudotsuga menziesii.

Leaf biomass of forests of Cryptomeria japonica, Pinus densiflora and Betula spp. is related to total above-ground biomass (Fig. 30). The leaf biomass increases with total biomass when the latter is small but becomes independent of total above-ground biomass when total biomass is large.

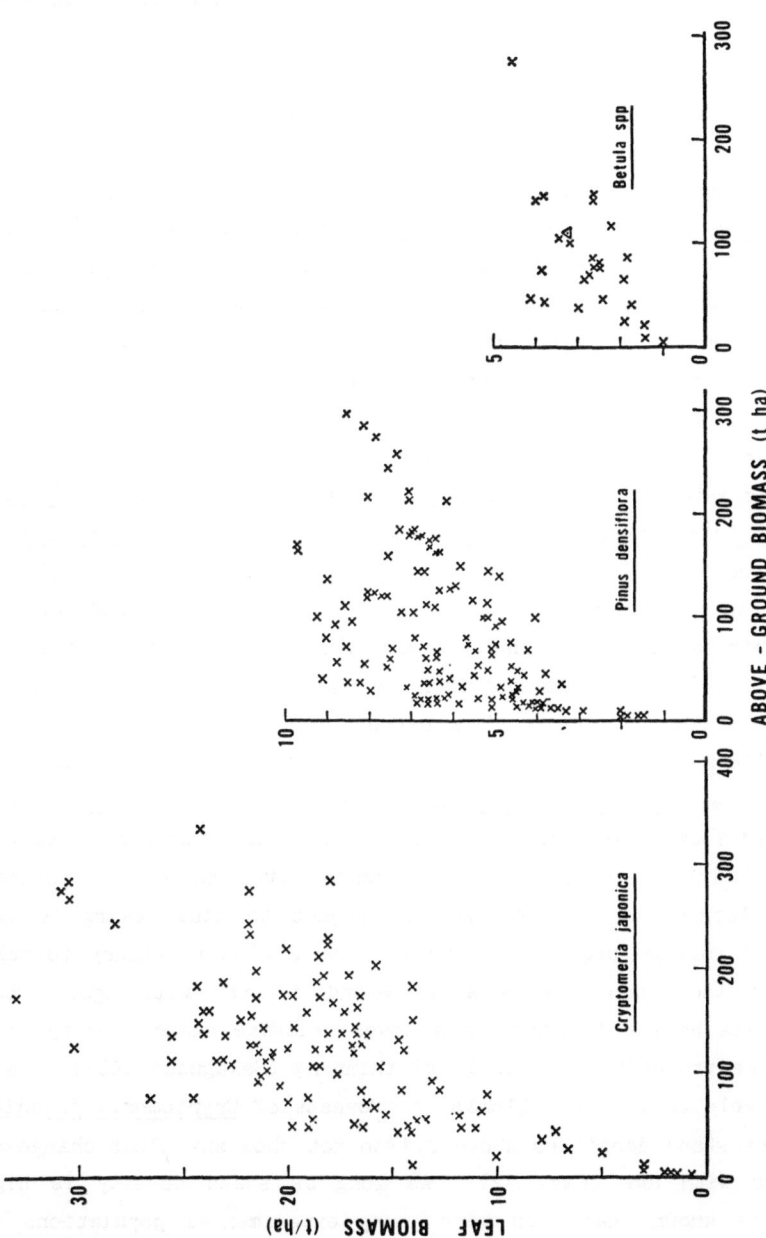

FIGURE 30. The relationships between total above-ground biomass and leaf biomass in stands of Cryptomeria japonica, Pinus densiflora and Betula spp.

4.1. Age

When a forest is clear-cut and replanted, or regenerates on open ground, leaf biomass starts at zero and increases more or less rapidly with age until the crown is closed (Table 13, Marks and Bormann 1972). The rate of increase depends on a variety of factors including species, site and stocking. For closed forests many different trends and patterns are reported. Möller (1945) reported that leaf biomass of forests of Fagus sylvatica and Picea abies increased slightly with age (Fig. 29), but according to Vanselow (1951) leaf biomass of Picea abies forests increased up to a maximum value at about the fortieth year and then decreased greatly. Kittredge (1944) reported that leaf biomass of Pinus ponderosa forests reached their maximum at the fortieth year and then gradually decreased, but in another table showing values of a similar nature, leaf biomass increased with age up to the sixty-ninth year. In forests of Pinus sylvestris (Table 13 and Utkin, et al. 1981), Abies veitchii-Abies mariesii (Oshima et al. 1958), Cryptomeria japonica (Ando et al. 1968), Pinus densiflora (Hatiya and Tochiaki 1968), and Pinus radiata (Forrest and Ovington 1970; Madgwick et al. 1977) leaf biomass increased until a maximum value was attained at an age dependent on the species and then decreased a little to maintain a steady state. Fig. 31 shows this pattern in a group of forests of Pinus densiflora as an example. Data from Betula spp. (Ovington and Madgwick, 1959a), Prunus pennsylvanica (Marks 1974) and Eucalyptus grandis (Bradstock, 1981) show a build-up to a maximum stable value. Whether the slight decrease in leaf biomass after an age of maximum foliage development is real or not is subject to doubt owing to the inaccuracy of foliage biomass estimates. It is also necessary to take into account the change of relative stand density with age. Most managed forests have a tendency to a lower relative stand density with increasing age probably as a result of thinning (Sakaguchi 1961). Data so far available on the leaf biomass of forests of Cryptomeria japonica with relative stand densities above 0.4 do not show any clear change of leaf biomass with age (Fig. 32). As long as stand density is high enough, it is known that even very young experimental populations of tree species have a leaf biomass comparable to mature forests (Satoo 1968b, Tadaki and Shidei 1960, Kawahara et al. 1968).

4.2. Site quality

Moller reported that among forests of Fagus sylvatica and Picea abies there was no difference in leaf biomass among forests on site qualities ranging from site index 1 to 4 in the Danish yield table (Fig. 29). However, Kittredge (1944) suggested that forests on better sites have larger leaf biomass. Data so far available on forests of Cryptomeria japonica with relative stand density above 0.4 show an increased foliar biomass with increasing relative height (Fig. 33, Satoo 1971b). Management practices for Cryptomeria forestry in Japan vary considerably among regions. Data from three regions having different management practices are illustrated in Fig. 34. In Yosino, trees are planted with very narrow spacing and thinned intensively. In Kumamoto, selected cultivars are propagated exclusively by means of cuttings and used as planting material. In Akita management practices are less intensive than in the other two regions. In each region leaf biomass increased with site index. Stands with relative stand density between 0.4 and 0.6 selected from data of Mori et al. (1969) showed a similar trend (Fig. 35). However, data of Hatiya et al. (1966) treated in the same way showed that leaf biomass increased with increasing site index among forests on poor sites but the increase was very slight among stands on better sites (Fig. 35). Among Picea abies stands planted in Japan, leaf biomass did not show any systematic trend with site index (Satoo, 1971a).

While there are conflicting conclusions based on existing information it seems that in general leaf biomass is larger on better sites.

4.3. Fertilization

There is an increasing tendency to modify the natural site quality of a forest by fertilization. Successful fertilization has been found to consistently increase foliar biomass for a variety of species including Pseudotsuga menziesii (Heilman and Gessel 1963) Pinus nigra (Miller and Miller 1976; Ranger 1978) and Pinus banksiana (Morrison and Foster 1977) fertilized with nitrogen, Pinus resinosa (Madgwick et al. 1970) fertilized with potassium, and Pinus thunbergii (Sakurai unpublished, Table 23). In Pinus resinosa the increase in total biomass was primarily due to an increase in needle longevity but

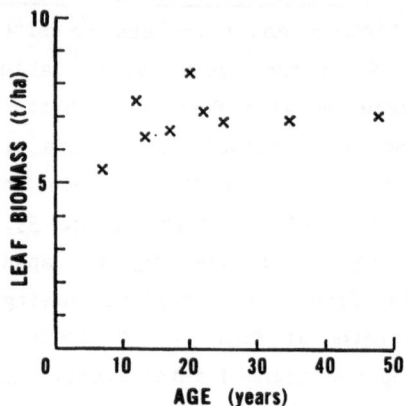

FIGURE 31. The relationship between stand age and leaf biomass of overstorey trees in forests of <u>Pinus densiflora</u> (Hatiya and Tochiaki 1968).

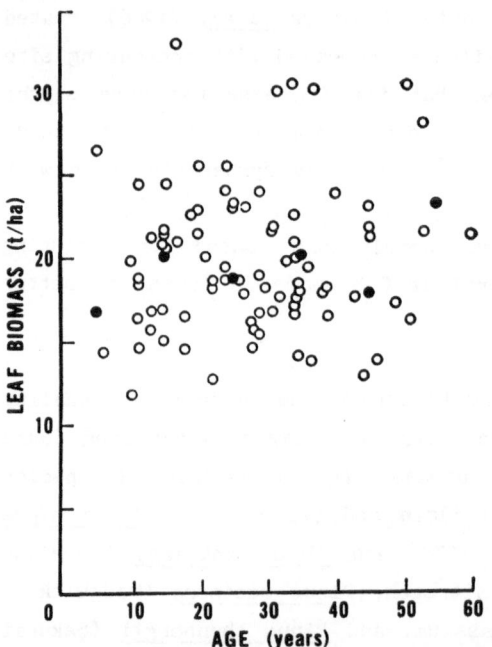

FIGURE 32. The relationship between stand age and leaf biomass of overstorey trees in forests of <u>Cryptomeria japonica</u> (Satoo 1971b). (o individual stands; ● mean values based on 10-year age classes).

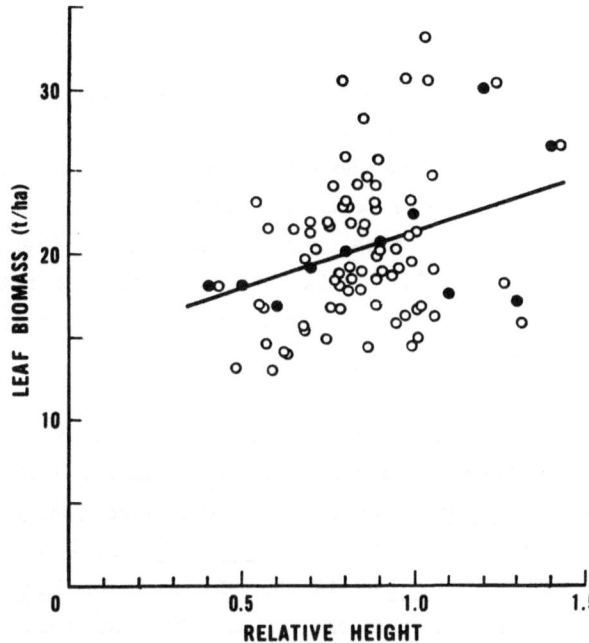

FIGURE 33. The relationship between site index expressed as relative height and leaf biomass of overstorey trees in plantations of <u>Cryptomeria japonica</u> based on stands with a relative stand density greater than 0.4 (Satoo 1971b). (o individual plots; ● mean values for relative height classes by 0.1 intervals; regression line based on individual plot data).

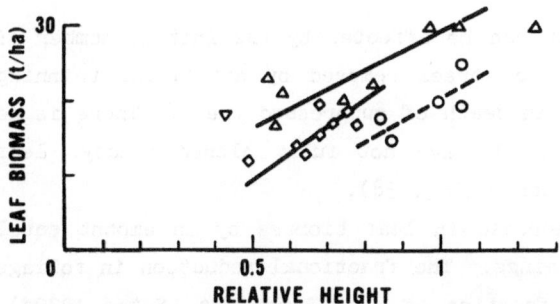

FIGURE 34. The relationship between site index expressed as relative height and leaf biomass of overstorey trees in plantations of <u>Cryptomeria japonica</u> by geographic region (Satoo 1966). (o Yosino; Akita; Kumamoto; solid regression lines are significant).

Brix (1981) has shown that with nitrogen fertilization of Pseudotsuga
menziesii needle longevity varies within tree crowns so making
comparisons complex.

4.4. Position on slope

It is often observed that soil properties and growth of trees
differ greatly between the top and bottom of a steep slope. On the
upper part of the slope a dry soil type develops and tree growth is
poor while on the lower part of the slope soils are moister with a rich
accumulation of organic matter and tree growth is good. These
differences can often be seen especially in plantations which have not
yet closed canopy. Fig. 36 shows an extreme example of a plantation of
very wide spacing in which this trend is very clear.

4.5. Altitude

Many factors affecting growth are severe at higher elevations and
leaf biomass of forests under these conditions is often small. This
trend has been reported for Picea abies forests of Switzerland (Burger
1942), for beech forests in Japan (Fig. 37) and Fagus - Betula - Acer
stands in North America (Whittaker et al. 1974). In contrast Pinus
pumila , a shrubby mountain species, has one of the highest foliar
weights reported for this genus (Madgwick et al., 1977). One of the
highest values of foliage biomass reported for Pinus contorta is from a
high elevation plantation in New Zealand (Nordmeyer 1980). Further
studies of elevation effects have been reported by Kimmins and Krumlik
(1973), Krumlik and Kimmins (1973) and Whittaker and Niering (1975) but
these are confounded with changes in species composition.

4.6. Stand density

Stand density of a forest can be affected by the initial number of
trees planted, by the number of trees removed by artificial thinning
and by competition resulting in death of suppressed trees. There is no
doubt that in young stands which have not fully closed canopy, leaf
biomass is larger in denser stands (Fig. 38).

Thinning results in a decrease in leaf biomass by an amount equal
to the leaf mass of the thinnings. The fractional reduction in foliage
mass will be similar to the fraction of basal area cut (Satoo 1979a).
Leaf mass then increases gradually to reach the level before thinning.
Möller (1945) reported that in Fagus sylvatica stands which had been

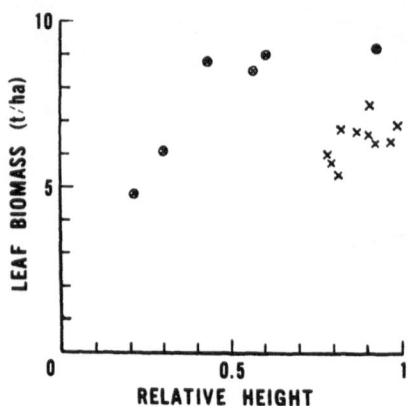

FIGURE 35. The relationship between site index expressed as relative height and leaf biomass of overstorey trees in forests of <u>Pinus densiflora</u> (o 21-year-old, relative stand density 0.85-1.19, Hatiya et al. 1966; x 55-year-old, relative stand density 0.43-0.60, Mori et al. 1969).

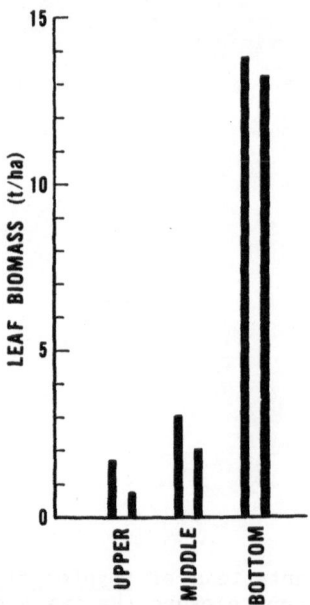

FIGURE 36. The effect of position on slope on the leaf mass of 20-year-old <u>Cryptomeria japonica</u> plantations on poor soils and at wide spacing (Hatiya and Ando 1965).

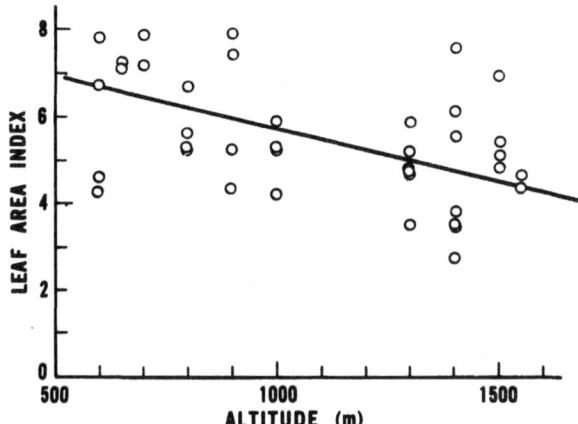

FIGURE 37. The change in foliage biomass with altitude in stands of
Fagus crenata (Maruyama 1971).

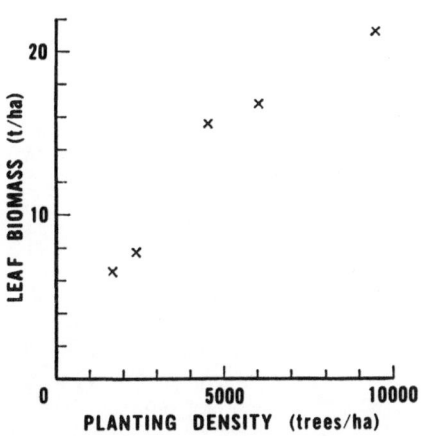

FIGURE 38. Foliage biomass on 13-year-old plantations of Cryptomeria
japonica as related to stand density before canopy closure (Hatiya and
Ando 1965).

differentially thinned to basal areas ranging from 17 to 35 square metres at age 50, leaf mass was unaffected by thinning intensity after recovery from the effects of thinning was complete (Fig. 39). Response to thinning was rapid in the first few years (Fig. 39). Eighteen years after plots of Pinus strobus in Japan had been thinned by removing 0 to 70 percent of their volume no difference in leaf biomass due to the intensity of thinning was observed (Fig. 40). However, where thinning is heavy or repeated, stands may not return to a foliage mass comparable to the unthinned condition. For instance, in stands of Liriodendron tulipifera with up to 75 percent of the basal area removed five years prior to determining foliage mass there was still a significantly lower leaf mass in heavily thinned than in lightly thinned stands (Madgwick and Olson, 1974). Similarly, Siemon (1973) found that repeated thinning of Pinus radiata plantations led to a close relationship between stocking and foliage mass.

Many spacing experiments show little effect of stocking on foliage mass once stands have closed. No difference was found in leaf biomass which could be attributed to differences in stand density in experiments on Pinus banksiana (Adams 1928), Pinus densiflora (Satoo et al. 1955, Senda et al. 1952) and Pinus sylvestris (Rubtsov and Rubtsov 1975; Zajaczkowski and Lech 1981). In very young experimental plots, 4 and 5 years old, leaf biomass was not different from closed 13-year-old stands provided that the young stands had a high stand density (Fig. 41).

A larger foliage biomass in heavily thinned stands with low stand densities might be inferred from their greater crown depth but this inference is incorrect since differences in crown depth only compensate for differences in spacing so that leaf biomass is similar in stands of very different densities having crown lengths ranging from 36 to 64 percent of height. However, there are also reports stating that leaf biomass increases with higher stand density. Sakaguchi et al. (1955) found this trend among young natural regeneration of Pinus densiflora. Leaf biomass is correlated with relative stand density in data from different sources, as shown in Fig. 42. A similar trend was found when existing data on the leaf biomass of Cryptomeria forests were treated in the same way (Fig. 43). Thus two apparently contradictory effects of density can be seen. Leaf biomass may be

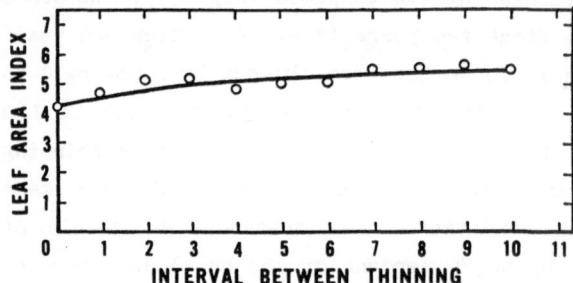

FIGURE 39. The change in leaf area index of <u>Fagus sylvatica</u> stands after thinning (Moller 1945). The time scale ranges from just after one thinning (0) to just before the next thinning (10).

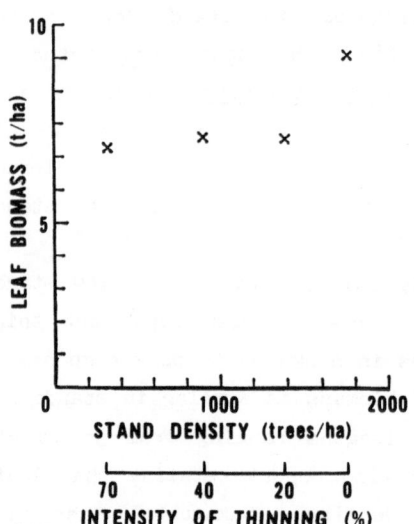

FIGURE 40. Leaf biomass on plantations of <u>Pinus strobus</u> 18 years after thinning (Senda and Satoo 1956).

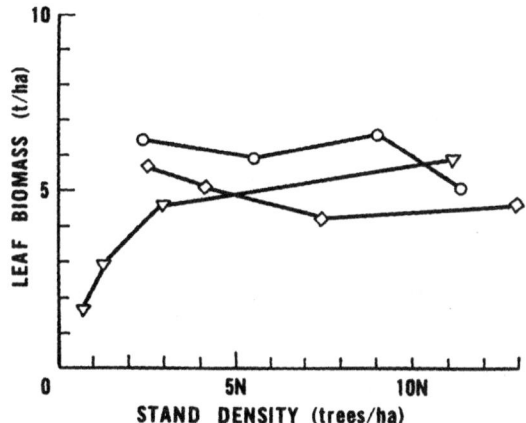

FIGURE 41. The relationship between leaf biomass and stand density in _Pinus densiflora_ (Satoo 1968). (o 4-year-old, N = 10^5; 5-year-old, N = 10^4; 13-year-old, N = 10^3).

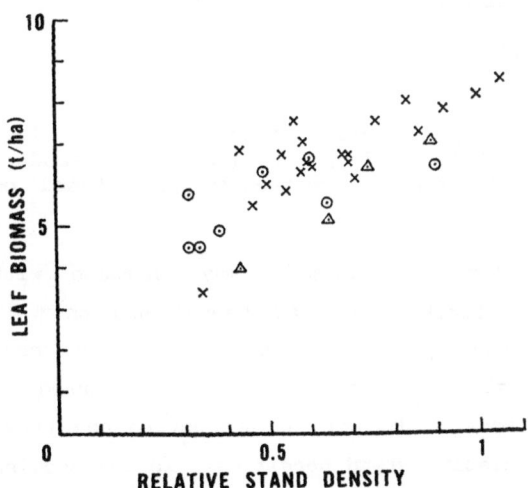

FIGURE 42. The relationship between relative stand density and leaf biomass in forests of _Pinus densiflora_ (46-year-old, Hatiya _et al._ 1965; o 14-year-old, Kato 1968; x 55-year-old, Mori _et al._ 1969).

76

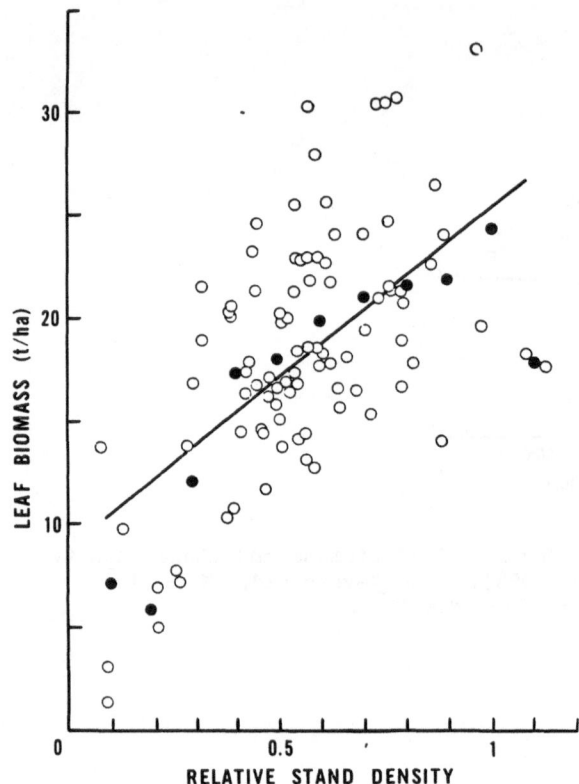

FIGURE 43. The relationship between relative stand density and leaf biomass in Cryptomeria japonica (Satoo 1971b). (o individual stands; ● mean values for 0.1 class intervals; regression line based on individual plot data).

independent of stand density or leaf biomass may increase with increasing stand density. The first trend has been found only in experimental plots with differences in spacing or thinning and these plots all have high relative stand densities. The second trend is found generally among forests managed for timber production and these forests have widely different relative stand densities. In the spacing experiment of Pinus densiflora (Satoo et al. 1955), for example, all plots had a relative stand density above 0.7. In the five-year-old experimental plots in Fig. 41, leaf biomass increased rapidly up to a stand density of 30 000/ha. Among managed forests of Cryptomeria (Fig. 43) leaf biomass increased very rapidly up to relative stand density of

0.4 above which the slope of the trend decreased. From these two apparently contradictory statements it may be said that leaf biomass is not affected by stand density as long as relative stand density is high but it increases with increasing stand density if relative stand density is low. In other words, leaf biomass is affected by relative stand density rather than by absolute stand density and managed forests frequently have comparatively low relative stand densities.

4.6. Silvicultural system

Burger (1942) reported that leaf biomass of a selection forest of spruce and fir did not differ from that of single-storeyed forest. The leaf area index was 10 in the selection forest compared with 11 in the single-storeyed forest. Selection forests may appear to have a larger biomass as their crowns are deeper as a result of the complexity of their crown structure.

Table 17. Leaf biomass of stands of Cryptomeria japonica in different regions of Japan (relative stand density over 0.4) (Satoo 1971b).

Region	No. of plots	Leaf biomass t/ha	
		Mean	Range
Tohoku, N.E. Honsyu	8	24.6	16-30
Kanto, near Tokyo	18	20.2	14-28
Kinki, near Kyoto	14	20.7	14-30
South-west Kyusyu	37	18.5	12-33
South-west Sikoku	5	20.4	17-22

4.8. Geographic region

Geographic variation in foliage mass was not found for Cryptomeria forests in Japan when stands with a relative density above 0.4 were compared (Table 17). In contrast, Pinus sylvestris growing over a wide range of climatic types in the Soviet Union has a foliage mass which is clearly related to region (Myakushko, 1974). Van Cleve et al. (1981) present evidence for a strong positive effect of temperature on the foliage mass of Picea mariana. The detection of geographic variation will undoubtedly be related to the range of climate within which a given species is growing.

4.9. Genetic variation

Genetic variation as determined by seed source was found not to affect foliar biomass of Picea abies (Burger, 1942). However, Pope (1979) compared four seed sources of Pinus taeda and found that the two sources with maximum foliage mass carried over twice the weight of needles on the source with lowest mass.

4.10. Year of sampling

Differences in foliage production of deciduous species from year to year have been reported to be negligible for Fagus sylvatica (Möller 1945) and Liriodendron tulipifera (Madgwick and Olson 1974). Year to year variation occurs in the development of Pinus sylvestris needles (Kishchenko 1978) and may influence total foliar biomass on evergreen species by affecting both needle production (Madgwick et al. 1970) and leaf fall.

4.11. Atmospheric pollution

Satoo (1979b) has found that foliage mass of Chamaecyparis obtusa influenced by atmospheric pollution from an electric power plant was only about 60 percent of that on a healthy stand in the same region.

4.12. Seasonal change

Obvious seasonal changes in foliar biomass occur in deciduous stands. This is illustrated for 3-year-old Ulmus parvifolia (Fig. 44) for which leaf mass increased until late May and then gradually decreased until leaf fall in the autumn. In contrast, Day and Monk (1977) working in a mixed deciduous forest dominated by Quercus prinus reported a gradual build-up of foliage till early June followed by a static foliar mass till early October. They found small differences among species in the time taken to reach peak foliar mass. The duration of leaf mass as determined by the date on which half the leaves had fallen off the trees was found to be highly correlated with site index and elevation in Liriodendron stands (Madgwick and Olson 1974).

Seasonal variations in foliage mass also occur in evergreen forests. Foliage production in a 20-year-old Pinus densiflora stand commenced in late spring (May) and was associated with the decline in the weights of older needles (Fig. 45). Some loss of current year needles occurred in late autumn so that the seasonal maximum foliage mass was 50 percent above the winter minimum (cf. Hatiya et al. 1966).

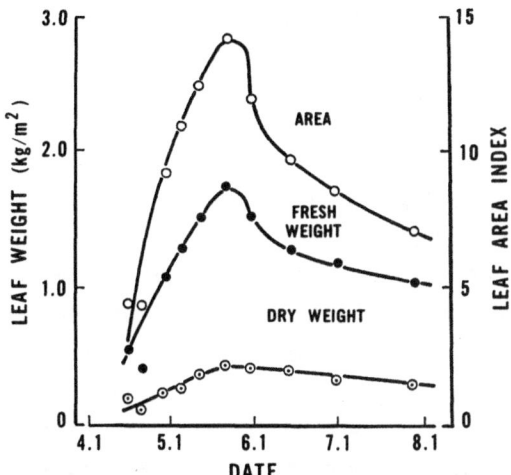

FIGURE 44. The seasonal change in leaf biomass in a miniature stand of Ulmus parvifolia (Tadaki and Shidei 1960).

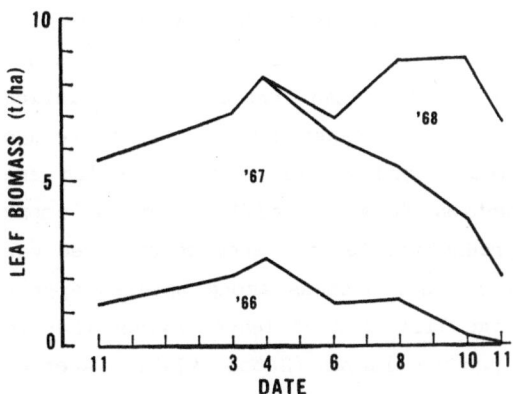

FIGURE 45. The seasonal change in leaf mass by needle age classes in a young stand of Pinus densiflora (Satoo 1971b).

A similar trend has been reported for <u>Pinus virginiana</u> (Madgwick 1968) while Rutter (1966) found even larger seasonal variations in foliar biomass of <u>Pinus sylvestris</u>. Early summer maximum foliage biomass has also been found for <u>Cryptomeria</u> (Fig. 46) and <u>Pinus nigra</u> (Miller and Miller 1976).

Leaf biomass data for forests of many species of trees in Japan are shown as dry weight and leaf area in Table 18 from Tadaki (1976). The values shown are mean values for existing data and could be modified as additional information is obtained. Leaf biomass differs not only by species, but also by the type of species. Species having similar ecological characters seem to have similar leaf biomass. Leaf biomass of <u>Betula</u> and <u>Fagus</u> of Japan and Europe are compared in Fig. 47. The data for each genus are similar for these widely separated geographical locations even though the species involved are different. The biomass of leaves on stands of conifers differs widely by species, with a maximum nearly eight times the minimum which is in strong contrast to the relatively narrow range of leaf biomass of deciduous broadleaved forests. Leaf biomass (F) can be characterized as the product of the average life span of leaf (Z) and the annual leaf production (ΔY_f), i.e.

$$F = \Delta Y_f \cdot Z$$

Z may be calculated by dividing F by ΔY_f since these values may be found for each stand. In Fig. 48 mean leaf biomass is plotted against both the mean life span of leaves and the mean annual leaf production for a range of species. Leaf biomass increases linearly with average life span, whereas total leaf biomass and annual leaf production are uncorrelated. Annual leaf production does not differ greatly among species and average annual leaf production for all species examined was 3.5 t/ha. The large differences of leaf biomass among species appear mainly due to differences in average life span of leaves rather than to any difference in the ability to produce leaves (Satoo 1971b). Average life span of leaves seems to be longer in more tolerant species (Monsi 1960). Tadaki (1966, 1976) summarised foliar mass for over 50 species. He considered the leaf biomass of deciduous forests to be the "basic unit of leaf biomass of closed forests" since the values obtained for such stands were broadly comparable to the annual production of foliage in evergreen stands.

Table 18. Leaf biomass of forest in Japan (Tadaki 1976).

Species	Leaf dry weight t/ha			Leaf area ha/ha		
	Mean	S.D.	No. of stands	Mean	S.D.	No. of stands
Deciduous broad leaved						
Fagus crenata	3.8	1.6	58	5.7	1.0	11
Betula ermanii	2.9	0.9	23	4.5	1.0	4
Betula platyphylla	1.1	0.3	13	2.8	0.7	12
Betula maximowicziana	2.3	0.5	3			
(Betula spp.)	2.2	1.1	39	3.2	1.1	16
Alnus-5 spp.	2.8	1.1	20	4.6	5.2	2
Populus-4 spp.	3.8	1.8	9	5.1	2.5	9
Others -11 spp.	2.4	1.0	15	5.4	1.9	10
Deciduous needle leaved						
Larix leptolepis	3.0	1.0	30	4.2		1
Metasequoia glyptostroboides	5.0		2			
Evergreen broad leaved						
Quercus phillyraeoides	8.6	2.0	7	6.7	1.7	7
Quercus glauca	6.3	0.5	3	7.1	0.3	3
Quercus myrsinaefolia	8.8	1.9	10	8.4	0.5	3
(Quercus spp.)	8.4	2.0	20	7.1	1.5	13
Castanopsis cuspidata	8.3	2.3	9	8.6	2.1	7
Machilus thunbergii	12.0	1.2	4	8.9	0.9	4
Camellia japonica	8.0	2.7	5	5.7	1.0	4
Acacia -2 spp.	5.7	2.3	11	7.1	2.0	6
Phyllostachys reticulata	6.8	0.6	3	11.3	1.7	3
Evergreen needle leaved						
Pinus densiflora	6.4	1.3	120			
Pinus thunbergii	7.7	2.1	15			
Pinus taeda	8.8	2.7	6			
Pinus elliottii	10.8	1.2	5			
Pinus strobus	6.1	2.5	6			
Pinus pumila	21.7	3.0	4			
Cryptomeria japonica	19.6	4.4	126	6.0	1.2	12
Chamaecyparis obtusa	14.0	2.5	26	5.1		2
Thujopsis dolabrata v. hondai	19.4	4.6	7			
Abies firma	16.5	2.6	4			
Abies veitchii (-A. mariesii)	16.1	4.7	38	9.9	1.9	7
Abies sachalinensis	22.8	3.7	5			
(Abies spp.)	16.8	4.9	47	9.9	1.9	7
Tsuga sieboldii	7.8		2			
Picea glehnii	17.0	6.3	5			
Picea abies	18.0	3.9	5			

5. REPRODUCTIVE STRUCTURES

The ephemeral nature of many reproductive organs has resulted in most studies of biomass being carried out when no such material is on the trees. In other studies the weights are small and are either neglected or included with some other component such as branches. A range of published values is included in Table 19. In pines the weight of male flowers is very variable with a maximum of 530 kg/ha reported for Pinus radiata. Cone production and retention varies among both species and individual trees. For instance, in a sample of 501 Pinus virginiana the smallest tree to bear cones had a diameter of 2.3 cm and the largest tree without cones 13.7 cm (Madgwick and Kreh 1980). Comparable figures for trees of Pinus radiata are 3.3 cm and 26.9 cm, respectively. In older stands of Pinus between 2 and 5 t/ha of cones may be found on the stand.

6. ROOTS

Excavating root systems is far more laborious than the measuring of above-ground biomass. In many experimental situations the disturbance caused by root extraction is unacceptable. In consequence simultaneous measurements of root systems and above-ground biomass are not common. Root studies are also plagued with a lack of standardisation. Jackson and Chittenden (1981) have shown that for Pinus radiata the relative amounts of root material in different size classes change rapidly, at least till trees are 10 m in height. Similarly, Ovington (1957) found that in Pinus sylvestris plantations roots less than 5 mm in diameter comprised 87 percent of total roots at age 7 years but only 37 percent at age 55 years (Table 13).

Following the summary of early studies reported by Bray (1962), root biomass has often been assumed to be between 20 and 25 percent of above-ground biomass. Fig. 49 shows a linear relationship between the biomass of above-ground and root systems measured on the same plots. The average ratios of root biomass to above-ground biomass for Pinus densiflora, Cryptomeria japonica and Abies veitchii were 0.24, 0.30 and 0.27, respectively. Santantonio et al. (1977) have tabulated results from over 100 non-Japanese sources. Individual variation in root:shoot ratios was very large ranging from 0.06 to 0.79 but averaged 0.30 for

Table 19. The weight of reproductive organs on tree stands (kg/ha)

Species	Male	Female	Reference
Chamaecyparis obtusa	–	120–160	Satoo (1979c)
Picea spp.	–	30–280	Gordon (1975)
Pinus banksiana	–	742–810	Morrison and Foster (1977)
Pinus radiata	277	158	Ovington et al. (1967)
	–	0–700	Forrest and Ovington (1970)
	530	0–5020	Madgwick et al. (1977)
Pinus sylvestris	–	0–1730	Ovington (1957)
Pinus virginiana	1–28	–	Madgwick and Kreh (1980)
Taxodium distichum	–	70	Schlesinger (1978)
Quercus rubra	–	20	Ovington et al. (1963)

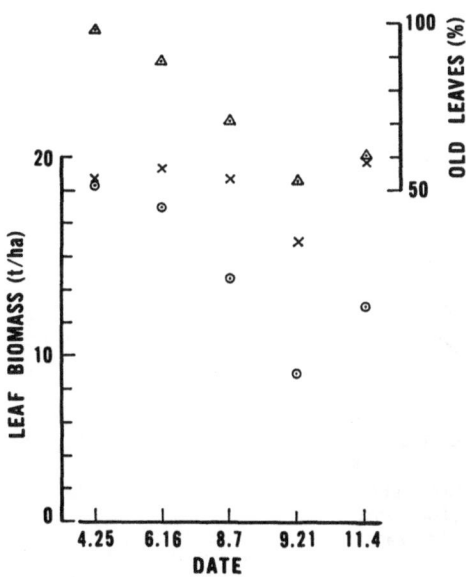

FIGURE 46. The seasonal change in leaf mass by age class and the changing percentage of old leaves in a plantation of Cryptomeria japonica (Ando and Takuchi 1968). (x total leaf; o old leaf; percentage of old leaves).

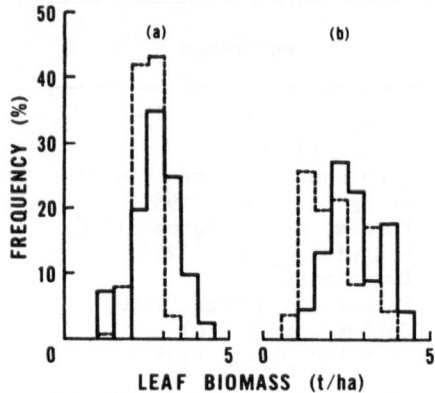

FIGURE 47. A comparison of leaf biomass of (a) <u>Fagus</u> and (b) <u>Betula</u> stands in Japan (solid lines) and Europe (broken lines) (Satoo 1970b).

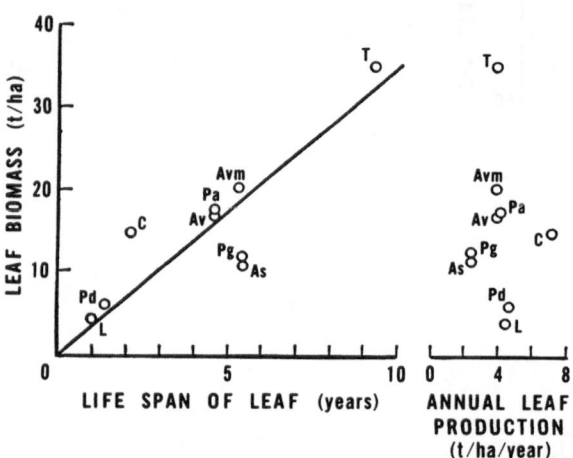

FIGURE 48. The relationships between leaf biomass and both life span and annual production of leaves (Satoo 1971b). (As, <u>Abies sachalinensis</u> (3 stands); Av, <u>Abies veitchii</u> (7); Avm, mixed <u>Abies veitchii</u> and <u>A. mariesii</u> (1); L, <u>Larix leptolepis</u> (4); Pg, <u>Picea glehnii</u> (1); Pd, <u>Pinus densiflora</u> (37); C, <u>Chamaecyparis obtusa</u> (1); T, <u>Thujopsis dolabrata</u> (3); Pa, <u>Picea abies</u> (6)).

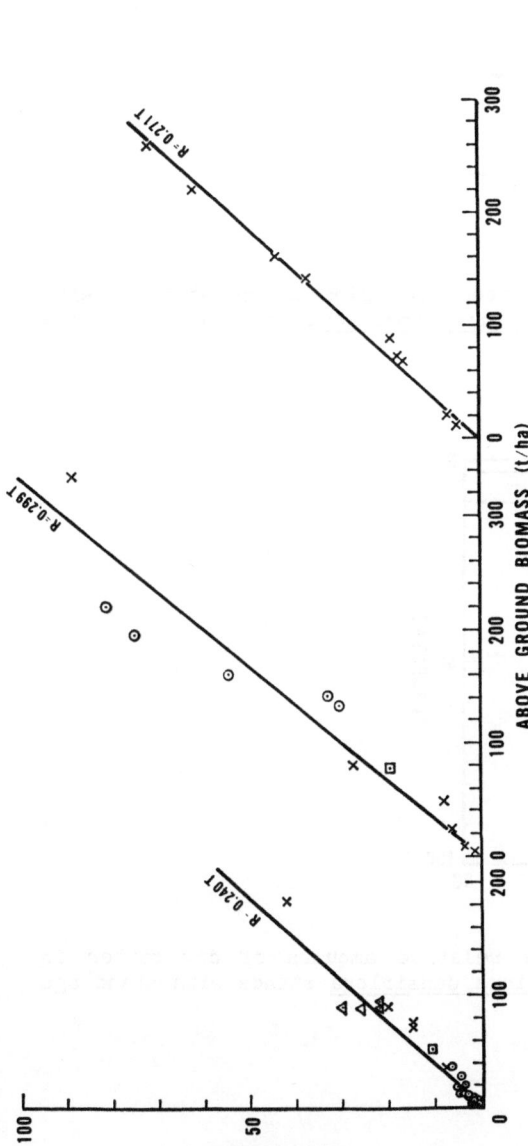

FIGURE 49. The relationship between root biomass and above-ground biomass in, from left to right, Pinus densiflora, Cryptomeria japonica and Abies veitchii.

conifers, 0.25 for temperate hardwoods and 0.25 for tropical forests. It does not appear satisfactory to assume a fixed root:shoot ratio for general use.

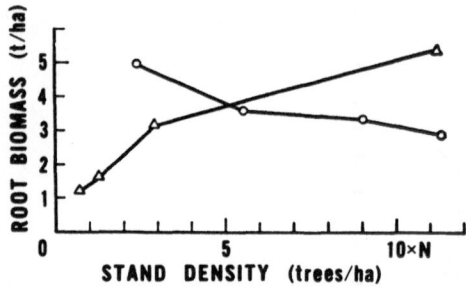

FIGURE 50. The relationship between root biomass and stand density in experimental plantations of <u>Pinus densiflora</u> (o 4-year-old, N = 10^5; ∆ 5-year-old, N = 10^4).

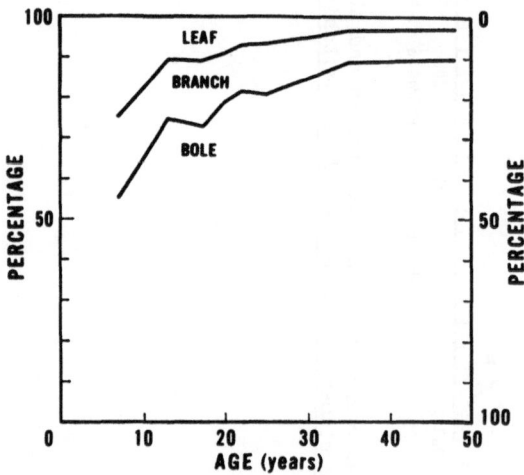

FIGURE 51. The change in the relative amounts of dry matter in leaves, branches and boles of <u>Pinus densiflora</u> stands with stand age (Hatiya and Tochiaki 1968).

As stands age there is a gradual build-up of roots above a diameter of 5 mm (Ovington 1957: Grier, et al. 1981) (Table 13). Smaller roots increase at a more rapid rate and then become more or less constant. The relationship between root biomass and stocking is less clear with different studies showing a negative, no or a positive relationship (Fig. 50; Baskerville 1966; Steinbeck and Nwoboshi 1980).

Fertilisation increases the biomass of small roots (Ranger 1978; Safford 1974) although comparisons of stands on good and poor sites have shown the reverse trend (Keyes and Grier 1981; Van Cleve et al. 1981).

A marked seasonal cycle occurs in the biomass of small roots though there is considerable variation in the pattern of change reported by different authors. Ovington et al. (1963) found a simple trend from a summer maximum to a winter minimum in a stand of Quercus rubra. Double maxima have been reported for Fagus sylvatica (Gottsche 1972 see Hermann 1977), Liriodendron tulipifera and Pinus taeda (Harris et al., 1977). For these species there were minima in mid-summer as well as in winter which agrees with the pattern found by Ovington et al. for prairie and savanna in the same locality as their Quercus forest.

7. RELATIVE DISTRIBUTION OF BIOMASS AMONG TREE COMPONENTS

The relative distribution of biomass among the parts of overstorey trees varies with conditions. Table 20 shows the changes in percentage of the main components of forest of Pinus densiflora, Cryptomeria japonica and Abies veitchii in relation to total biomass. The percentage in boles increases with the increase of total biomass as production is accumulated year by year. Leaf production is poorly related to total biomass (Fig. 30) so the percentage of leaves decreases with increased biomass. The percentage of branch material decreases with increase of total biomass in the case of Pinus densiflora but not for either of the other two species. The relation of the percentage of root biomass to total biomass is not clear for these three species. Since information on root biomass is limited, only data for above-ground parts have been plotted against various factors. The percentage of bole biomass increases and the percentages of branch and leaf biomass decrease with increasing age (Fig. 51), site index (Fig. 52) and stand density (Fig. 53).

Table 20. The percentage of biomass in components of tree stands by classes of total above-ground biomass

Species		Stand biomass t/ha					
		0-25	25-50	50-100	100-200	200-400	Over 400
Pinus densiflora	No. of samples	11	4	3	5	1	0
	Leaves	30	21	7	6	3	–
	Branches	20	21	12	11	6	–
	Boles	31	41	65	62	72	–
	Roots	18	17	17	21	19	–
Cryptomeria japonica	No. of samples	2	1	2	3	3	1
	Leaves	32	17	20	11	8	6
	Branches	6	4	4	6	7	4
	Boles	39	58	60	63	59	69
	Roots	23	21	16	21	26	21
Abies veitchii	No. of samples	1	1	2	2	3	0
	Leaves	33	28	18	15	6	–
	Branches	11	13	13	13	8	–
	Boles	30	36	50	52	64	–
	Roots	26	23	19	20	22	–

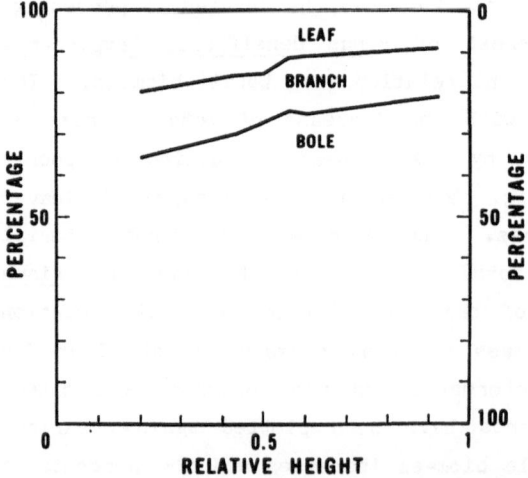

FIGURE 52. The change in relative amounts of foliage, branches and boles of 21-year-old Pinus densiflora stands in relation to site index expressed as relative height (Hatiya et al. 1966a).

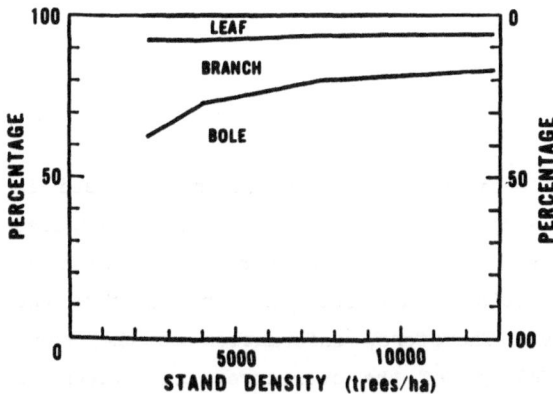

FIGURE 53. The change in relative amounts of foliage, branches and boles of 13-year-old plantations of Pinus densiflora in relation to stand density (Satoo et al. 1955).

The observed effects of a wide variety of stand and environmental variables on components of forest biomass suggest that published information on biomass should be accompanied by full descriptions of stand conditions. Some authors have failed to give such basic information as basal area even though the necessary information was required to calculate stand biomass. A minimal set of data should include location, age, basal area, stand height, numbers of stems and season of sampling. Such information would materially increase the possibility of developing general relationships for describing forest biomass.

5. PRODUCTION

1. NET PRODUCTION

Among the many methods of estimating net production described in Chapter 3 only the results using the harvest method are discussed here. This represents a fairly large accumulation of data. Most of the work on the dry matter production of forests has been done from the standpoint of applied forestry. Therefore, information on the stems and to a lesser extent, the foliage of the overstorey tree layer of forest ecosystems is abundant, while measurements of production by other components such as branches, roots and understorey trees and undergrowth at the same time as overstorey tree layers is limited. Moreover, the difficulties associated with estimating death and shedding of branches and roots are such that accurate estimates of their production are few in number and doubtful in absolute value. Table 21 gives an example of a detailed measurement of net production by layers for a planted larch forest. Table 22 shows the relationships between above-ground production of the overstorey tree layer and undergrowth of many different types of forests. The proportion between the two components is very variable.

2. UNDERGROWTH

Net production by undergrowth decreases with increasing leaf biomass of the overstorey. As shown in Fig. 54, the logarithm of net production by undergrowth (Pn) decreases linearly with the increase of leaf area index (F) of the overstorey tree layers, and may be expressed as

$$\log Pn = - aF + b \quad\quad\quad\quad\quad (5.1)$$

where a and b are constants. As described in Chapter 2

$$F = -1/K \log \frac{I}{Io} \quad\quad\quad\quad\quad (2.2)$$

Substituting for F in equation (5.1) we get

$$\log P_n = \frac{a}{K} \log \frac{I}{I_o} + b \quad \dots\dots\dots\dots\dots\dots\dots\dots\dots\dots \quad (5.2)$$

suggesting that the logarithm of net production by undergrowth is linearly related to the logarithm of relative light intensity. The curve showing the relationship between production and relative light intensity is nearly linear within the range of lower light intensity (Monsi and Saeki 1953).

Table 21. Net production in a 39-year-old plantation of <u>Larix</u> <u>leptolepis</u> (t/ha/year) (Satoo 1970a)

	Overstorey larch trees	Second storey deciduous broadleaved trees	Shrubs	Ground vegetation	Total
Leaves	3.59	0.31	0.19	0.36	4.45
Woody parts	9.06	0.57	0.41	-	-
bole	5.80	0.35	-	-	-
branch	3.26	0.22	-	-	-
Above-ground	12.64	0.88	0.60	-	-
Total	-	-	-	1.70	-

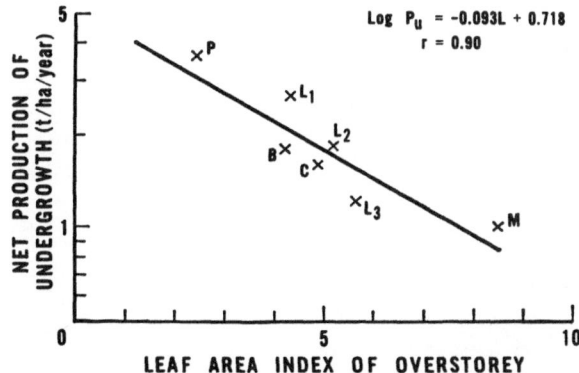

FIGURE 54. The relationship between net annual production of understorey (Pn) and the leaf area index of the overstorey (F). (P = <u>Populus davidiana</u> forest; B = <u>Betula ermanii</u> forest; C = <u>Cinnamomum camphora</u> plantation; M = <u>Metasequoia glyptostroboides</u> plantation: L = <u>Larix leptolepis</u>, 1 below second storey, 2 below shrub layer, 3 below ground vegetation)

Table 22. Net production of overstorey and undergrowth

Species of overstorey trees	Net production (t/ha/yr)				Source
	Over-storey trees	Under-growth	Total	Under-growth as % of total	
Betula ermanii	7.6	1.4	9.0	16	Tadaki & Hatiya 1970
Populus davidiana	8.7	3.6	12.3	30	Satoo et al. 1956
Cinnamomum camphora	13.6	1.6	15.2	11	Satoo 1968a
Larix leptolepis	14.5	0.6	15.1	4	Satoo 1974b
Larix leptolepis	12.6	2.7	15.3	18	Satoo 1970a
Metasequoia glyptostroboides	16.2	1.0	17.2	6	Satoo 1974d
Abies sachalinensis	14.5	(+)	14.5	(+)	Satoo 1974c
Abies sachalinensis	12.5	(+)	12.5	(+)	4 universities data
Picea abies	11.7	1.4	13.1	11	Satoo 1971a
Picea abies	12.4	1.6	14.0	12	Satoo 1971a
Picea abies	11.4	1.4	12.8	11	Satoo 1971a
Picea abies	7.3	1.4	8.7	16	Satoo 1971a
Picea glehnii	7.4	(+)	7.4	(+)	4 universities data
Thujopsis dolabrata	11.8	(+)	11.8	(+)	Satoo et al. 1974
Thujopsis dolabrata	19.2	(+)	19.2	(+)	Satoo et al. 1974
Thujopsis dolabrata	13.0	(+)	13.0	(+)	Satoo et al. 1974
Pinus densiflora	14.5	(+)	14.5	(+)	Satoo 1968b

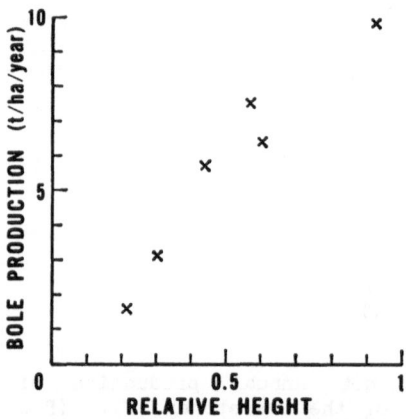

FIGURE 55. The relationship between bole production and site index expressed as relative height in 21-year-old forests of Pinus densiflora (Hatiya et al. 1966)

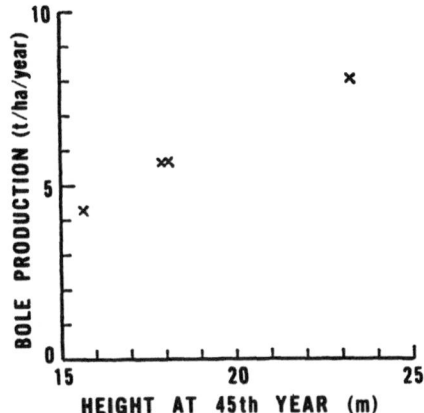

FIGURE 56. The relationship between bole production and site index expressed as height at age 45 years in 45- to 47-year-old plantations of Picea abies (Satoo 1971a).

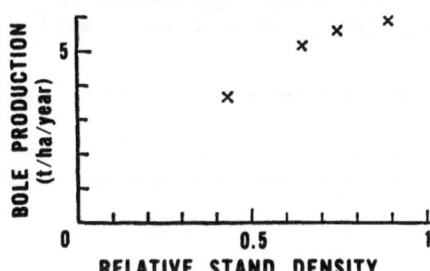

FIGURE 57. The relationship between bole production and relative stand density in 43- to 46-year-old forests of Pinus densiflora (Hatiya et al. 1965).

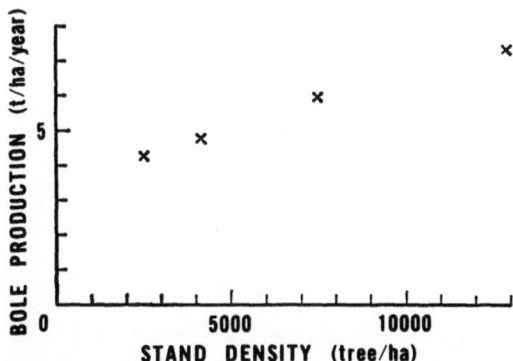

FIGURE 58. The relationship between bole production and stand density in plantations of Pinus densiflora of very high stand density (Satoo et al. 1955)

94

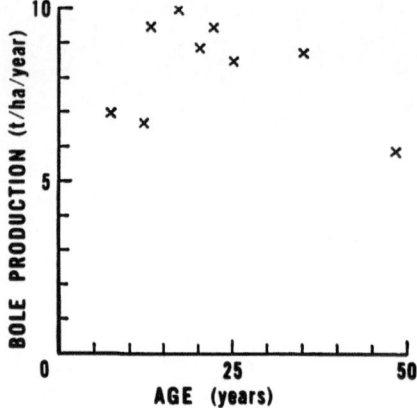

FIGURE 59. The relationship between bole production and stand age in forests of <u>Pinus densiflora</u> (Hatiya and Tochiaki 1968).

Table 23. Change of net production by heavy application of fertiliser in a plantation of <u>Pinus thunbergii</u> on a coastal sand dune (Sakurai, unpublished)

	Unfertilised	Fertilised
Leaf biomass t/ha	6.96	7.85
Net production t/ha/yr		
leaves	3.99	4.65
branches	1.70	2.05
boles	3.09	7.50
Total above-ground	8.78	14.20

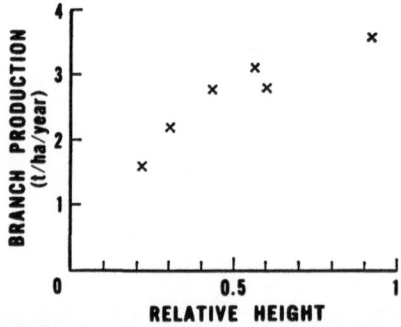

FIGURE 60. The relationship between branch production and site index expressed as relative height for 20- to 21-year-old forests of <u>Pinus densiflora</u> (Hatiya <u>et al</u>. 1966).

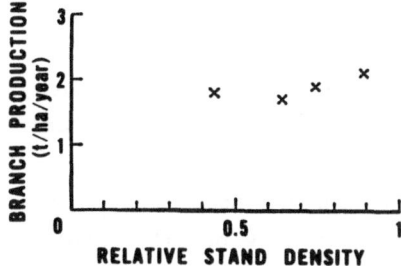

FIGURE 61. The relationship between branch production and relative
stand density for 43- to 46-year-old forests of _Pinus densiflora_
(Hatiya _et al_. 1965).

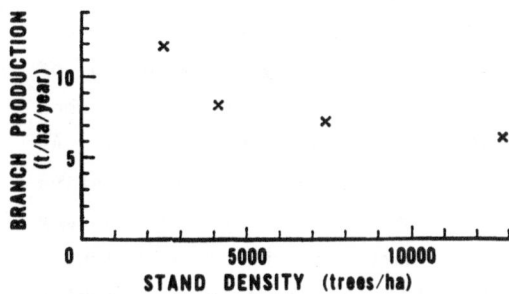

FIGURE 62. The relationship between branch production and stand
density in 15-year-old plantations of _Pinus densiflora_ with very
high stand densities.

3. OVERSTOREY TREES

3.1. Bole

Bole production has been the direct objective of forestry practice
for many decades so abundant information has been accumulated for
predicting the yield of wood in forestry. Though this information is
expressed as volume, not as dry weight, the average density of wood of
important species is available in many handbooks and, using these
values, volume can be converted into dry weight. The production of
bole is dependent on many factors. Bole production increases with

increasing site index (Figs. 55 and 56) either expressed as relative
or absolute height. Total bole production increases also with
relative stand density (Fig. 57) and with absolute stand density
even among forests having high relative stand density (Fig. 58)
although the size of individual trees decreases with increasing
absolute and relative density. Bole production at first increases
with age to a maximum and then gradually decreases as shown by Fig.
59. The age at which current annual growth of bole becomes a
maximum varies with species, stand density and site quality. Bole
production may also be increased by application of fertilizer as
illustrated for sand dune plantations of Pinus thunbergii (Table 23).

3.2. Branches

Systematic utilization of branches began only recently with
their use as a source of fibre. The measurement of production of
branchwood is laborious and information on branch production is
limited. Branch production is known to be affected by various
factors. It sometimes increases with increasing site index as shown
in Fig. 60, and it may increase after heavy application of
fertilizer (Table 23 and Madgwick 1975). There are other cases when
branch production is not influenced by site index (Satoo 1971). As
opposed to bole, branch production is unaffected by either relative
or absolute stand density (Fig. 61 and Madgwick 1970b) but may
decrease with increasing stand density among forests of sufficiently
high relative stand density (Fig. 62) or where high stocking
decreases stand growth rates (Madgwick 1975). Branch production by
individual trees decreases with increased stand density. Production
by individual branches is closely related to branch dimension (Fig.
9 and 10), regardless of the height in the canopy at which the
branch exists. Branch production in a layer within the crown canopy
is closely related to the leaf biomass in the layer regardless of
the position of the layer within the canopy (Fig. 63 and Madgwick
1974).

3.3. Leaves

Leaf production increases with site index, approaching an upper
limit asymptotically at high site indices (Fig. 64). Leaf
production increased after the heavy application of fertilizer on a
plantation of Pinus thunbergii on a sand dune site (Table 23) but

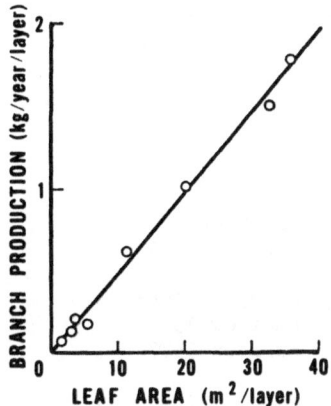

FIGURE 63. The relationship between branch production and leaf area of 2-metre deep layers in the canopy of a 14-year-old <u>Metasequoia glyptostroboides</u> plantation (Satoo 1974d).

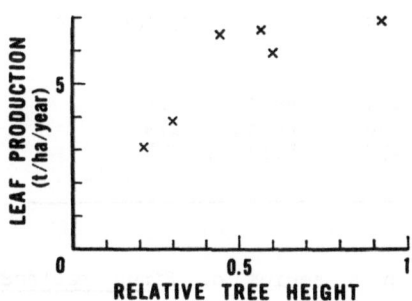

FIGURE 64. The relationship between leaf production and site index expressed as relative height for <u>Pinus densiflora</u> overstorey (Hatiya et al. 1966).

Table 24. Estimates of annual production of reproductive structures

Species	Structure	Weight kg/ha	Source
Alnus glutinosa -	Male	88	Hughes 1971
Betula pendula	Female	58	Hughes 1971
Alnus rubra	Catkins - cones	6-1080	Zavitkovski and Newton 1971
Betula spp.	Seed	283	Ovington 1963
Carpinus betulus -	Flowers	17- 440	Kubicek 1977
Quercus spp.	Seeds	51-1097	Kubicek 1977
Chamaecyparis	Cones	56- 128	Satoo 1979c
obtusa	Seed	17- 40	Satoo 1979c
	Male flowers	3- 32	Satoo 1979c
	Sexual organs	7- 380	Hagihara 1978
Pinus strobus	Male cones	188	Ovington 1963
Populus tremuloides	Female catkins	543	Ovington 1963
Quercus petraea	Male inflorescences	28	Ovington 1963
	Acorns	0- 645	Ovington and Murray 1964
Quercus robur - Tilia cordata	Inflorescences	75	Bandola-Ciolczyk 1974
Thuja occidentalis -	Thuja cones	269	Reiners 1974
Betula papyrifera	Thuja seeds	58	Reiners 1974
	Betula seeds and scales	477	Reiners 1974

was independent of potassium status in a series of Pinus resinosa stands with differing height growth rates on outwash sands several years after fertilization (Madgwick 1975). The effects of stand density are ambiguous. At low densities, leaf production of Pinus densiflora increased with increasing relative stand density (Fig. 65), at high relative stand density leaf production was independent of relative stand density (Fig. 66) but in Pinus resinosa where high density was associated with a decrease in stand growth there was a negative relationship between leaf production and density (Madgwick et al. 1970). Leaf production usually increases with stand age, reaching a maximum about the time of stand closure after which a slight decline occurs (Fig. 67). In _Pinus radiata_ this decline was associated with an increase in needle longevity (Madgwick _et al_. 1977).

3.4. Reproductive structures

Information about the production of these minor components is scarce. However, the percentage by weight of total net production in reproductive organs is usually small. For example, of the 15.4 t/ha of above-ground net production in a stand of _Chamaecyparis obtusa_, production of seeds and cones was only 0.12 t/ha (Tadaki _et al_. 1966). Estimates of the annual production of ephemeral structures such as strobili can be obtained from biomass data collected at the relevant time of year (Table 19). Other estimates based on litter fall are given in Table 24. The production of reproductive structures varies greatly from year to year (Ovington and Murray 1964; Zavitkovski and Newton 1971; Kubicek 1977). In _Alnus rubra_ there was little production for the first 5 years while from age 11 to 34 years between - stand variation obscured any age effect (Zavitkovski and Newton 1971).

3.5. Roots

Information about root production is limited because of the difficulties of measurement. Methods of estimating root production are not well established. Of 15.8 t/ha of total production by a 15-year-old forest of _Pinus densiflora_, 1.3 t/ha was root production (Satoo 1968b). Of 17.5 t/ha of total production by a 30-year-old plantation of _Chamaecyparis obtusa_ root production was 2.3 t/ha (Yamakura _et al_. 1972). Harris _et al_. (1977) have estimated annual rootlet production from the seasonal changes in standing crop. They estimated a production of 9.0 and 8.6 t/ha/annum for _Liriodendron tulipifera_ and _Pinus taeda_, respectively. Applying their method to the data of Ovington _et al_. (1963) yields an estimated production of 7.7 tonnes from April to July in a _Quercus_ forest.

3.6. Above-ground net production

The production of bole, branches and leaves combined make up most of the above-ground production. If we could include production of bark, seeds, fruits, premature leaf fall, etc., the value of above-ground production would be a little larger, and the values published so far must be slight underestimates. However these biases are not large

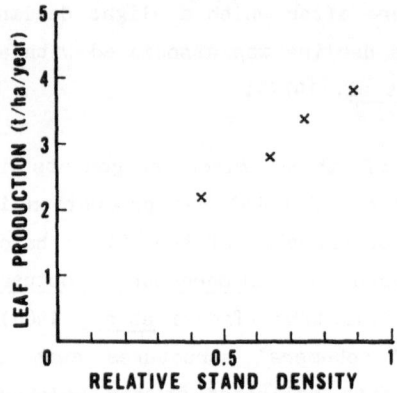

FIGURE 65. The relationship between leaf production and relative stand density in 43- to 46-year-old forests of _Pinus densiflora_ (Hatiya _et al._ 1965).

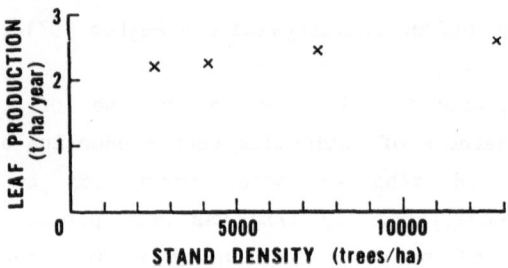

FIGURE 66. The relationship between leaf production and stand density in a 13-year-old plantation of _Pinus densiflora_ with very high stand density (Satoo _et al._ 1955).

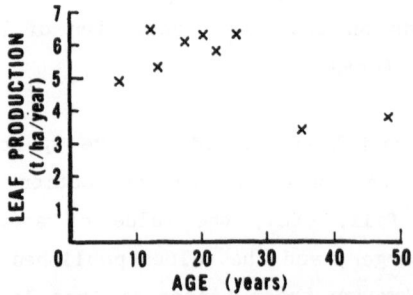

FIGURE 67. The relationship between leaf production and stand age in forests of _Pinus densiflora_ (Hatiya and Tochiaki 1968).

when we consider the errors involved in measurement. Figs. 68 and 69
show the relationship between above-ground net production and site
index of Japanese plantations of Pinus densiflora and Picea abies,
respectively. In both species, above-ground net production increased
with increasing site index. Fertilization also increased total
production on coastal sand dune plantations of Pinus thunbergii (Table
23). Above-ground production increased with both relative stand
density (Fig. 70) and absolute stand density among forests of
sufficiently high relative stand density (Fig. 71). Above-ground
production increased with age up to a certain age and then decreased
(Fig. 72). It also decreased with increasing altitude (Fig. 73).

Existing data on the above-ground net production of the overstorey
tree layer of forests in Japan are summarized in Fig. 74. Production
by deciduous broadleaved forests is smaller than other forest types
among which differences are not great. In Table 25 are shown the
values of above-ground production by different forest types within a
few kilometres of each other in the Tokyo University Forest in
Hokkaido. Except for a Picea abies plantation on a very poor soil,
conifer plantations produced more dry matter than second growth
deciduous broadleaved forests. Ovington (1956) has also reported on
the comparative net production of a wide variety of species on three
different sites in England and found similar results.

Table 25. Net above-ground production of different forest types in
Tokyo University Forest in Hokkaido all located within a radius of a
few kilometers (Satoo 1970)

Species	No. of stands	Age years	Net production t/ha
Betula maximowiziana	3	47	4.2-6.2
Populus davidiana	1	40	8.7
Picea abies	4	45-47	7.3-12.4
Abies sachalinensis	1	26	13.0
Larix leptolepis	1	21	16.5

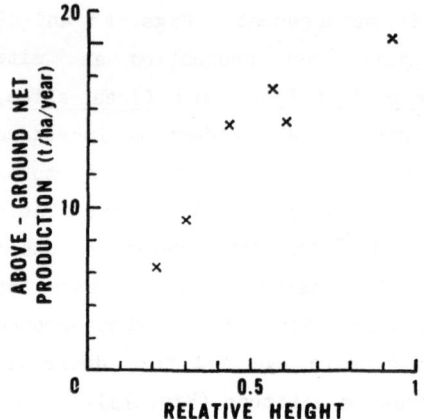

FIGURE 68. The relationship between above-ground net production by the overstorey of 20- to 21-year-old forests of <u>Pinus densiflora</u> and site index expressed as relative height (Hatiya <u>et al</u>. 1966).

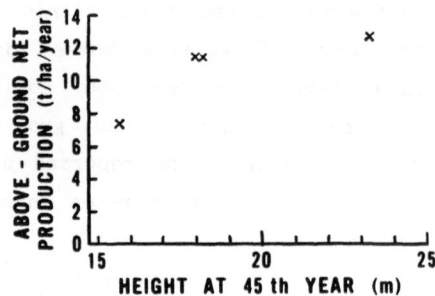

FIGURE 69. The relationship between above-ground net production by overstorey <u>Picea abies</u> stands and site index expressed as height at age 45 years (Satoo 1971a).

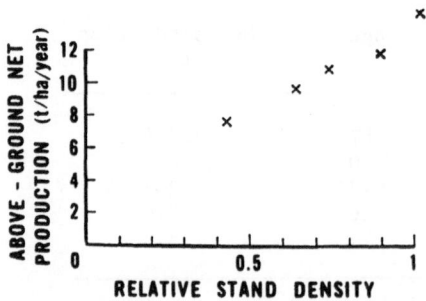

FIGURE 70. The relationship between net above-ground production by 43- to 46-year-old <u>Pinus densiflora</u> stands and relative stand density (Hatiya <u>et al</u>. 1965).

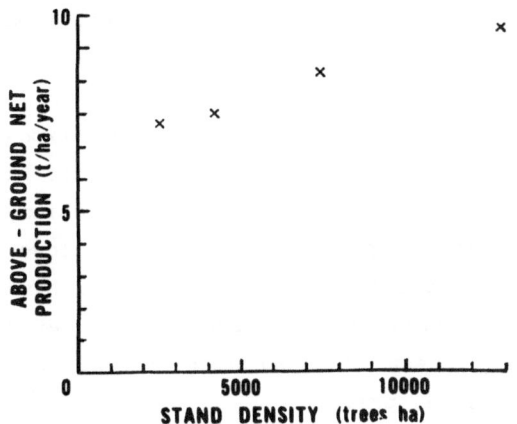

FIGURE 71. The relationship between above-ground net production and stand density of a 15-year-old <u>Pinus densiflora</u> plantation of high stand density.

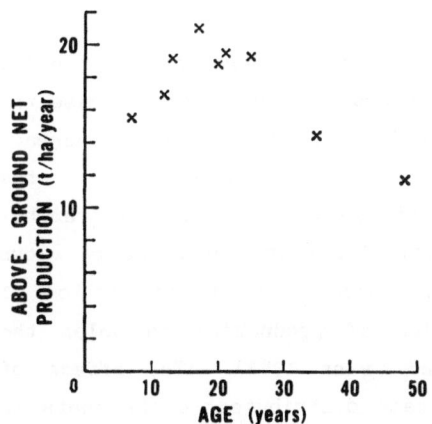

FIGURE 72. The relationship between above-ground net production and stand age in forests of <u>Pinus densiflora</u> (Hatiya and Tochiaki 1968).

The annual net production of agricultural crops was estimated from mean yields and the ratio of yield to net production calculated from published data for comparison with tree data in Table 25. Production of beets and potatoes was 11 and 8.5 t/ha/year, respectively which is lower than the above-ground net production of coniferous plantations (Satoo 1970). Although agricultural crops are grown on better soils and more intensively managed using techniques such as cultivation, fertilization and breeding, their annual production is usually lower than that of either evergreen or deciduous forests as the latter have well-developed foliage soon after the growing season becomes favourable for photosynthesis. Consequently forests use solar radiation more efficiently when considered on an annual basis.

3.7. <u>Distribution of net production among tree components</u>

The distribution of dry matter production plays an important role in utilization of the products of photosynthesis and in the flow of material and energy in forest ecosystems. For the overstorey tree layer, production of harvestable bole (Pb) depends not only on total net production (Pn) but also on the fraction (f) distributed to boles.

$$Pb = Pn \cdot f \quad \dotsfill \quad (5.3)$$

Consequently there are two ways of increasing the production of bole. One is to increase net production itself and another is to increase the fraction of production distributed to bole. Up to a point an increase in the fraction distributed into bole results in increased total biomass, since bole material does not die as early as either branches or leaves, thus turnover rates of both dry matter and energy slows down. Moreover, boles remain the most commonly harvested portion of the forest so the larger the fraction of production in boles the greater the usable yield (Gifford and Evans 1981). The effect of changes in the proportion of photosynthate distributed to the roots is not clear.

The pattern of distribution is affected by various factors. As shown in Fig. 75, the distribution into boles increases and into leaves and branches decreases with increasing site index, though an exception has been noted (Satoo 1971). With fertilization there may be a relative shift from leaves to branches (Madgwick 1975) and from boles to branches (Will and Hodgkiss 1977). The fraction distributed to

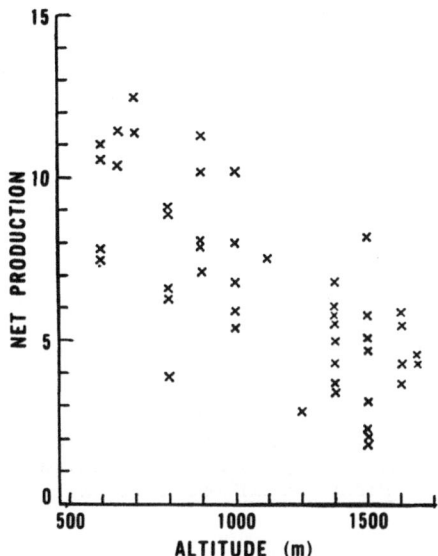

FIGURE 73. The relationship between above-ground net production of natural stands of *Fagus crenata* and altitude (Maruyama 1971).

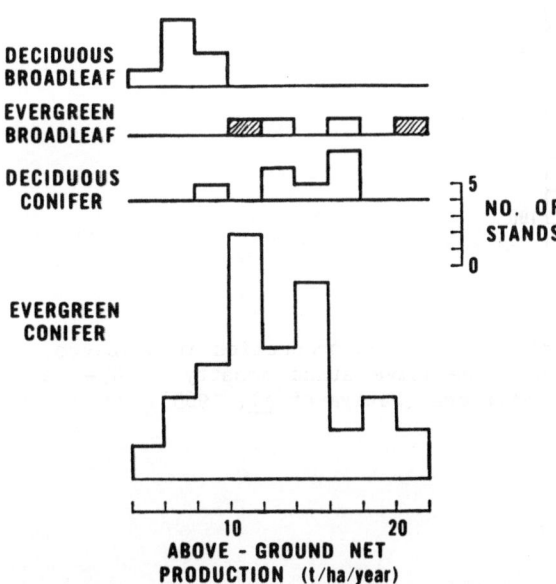

FIGURE 74. The above-ground net production of forests of different types in Japan (hatched = *Acacia*).

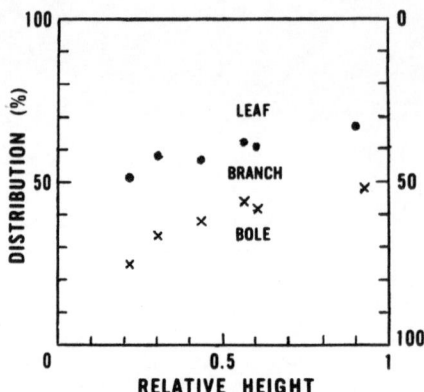

FIGURE 75. The relative distribution of net production to leaves, branches and boles in relation to site index expressed as relative height for 20- to 21-year-old forests of Pinus densiflora (Hatiya et al. 1966).

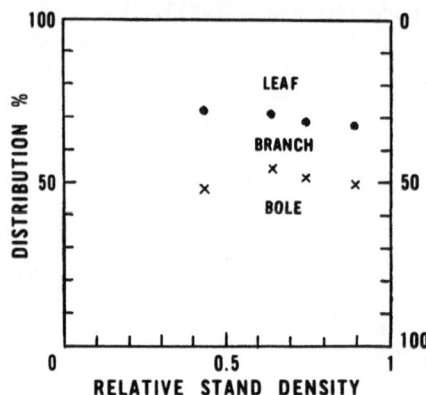

FIGURE 76. The relative distribution of net production into leaves, branches and boles in relation to relative stand density in 43- to 46-year-old forests of Pinus densiflora (Hatiya et al. 1965).

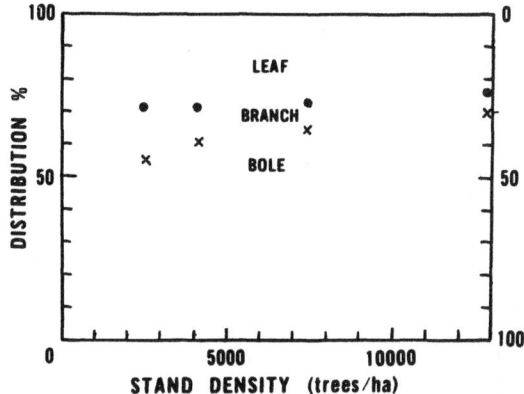

FIGURE 77. The relative distribution of net production in leaves, branches and boles in relation to stand density in a _Pinus densiflora_ plantation of high stand density.

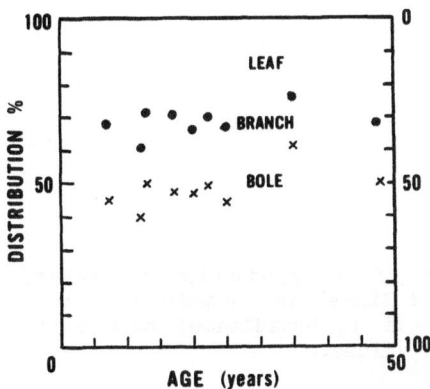

FIGURE 78. The relative distribution of net production in leaves, branches and boles in relation to stand age for forests of _Pinus densiflora_ (Hatiya and Tochiaki 1968).

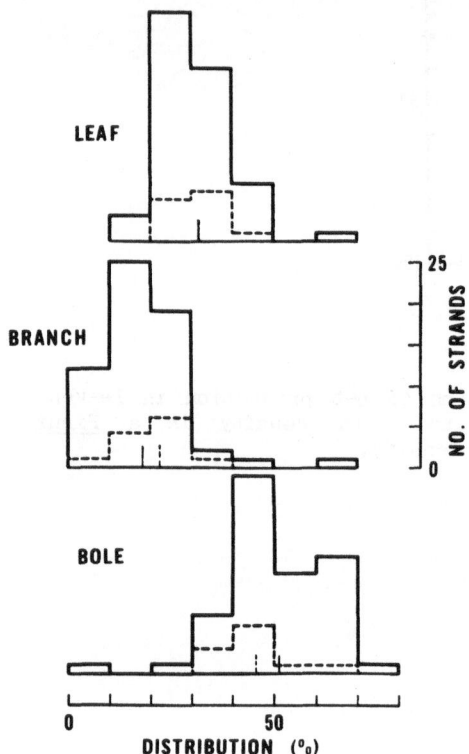

FIGURE 79. The relative distribution of net production in leaves, branches and boles for conifers (solid lines) and broadleaved trees (broken lines) based on 60 conifer and 12 broadleaved data sets. The short vertical lines indicate mean values.

boles does not seem to change with relative stand density at low densities (Fig. 76) but increases, at the expense of branches, with increasing stand density among forests with sufficiently high relative stand density (Fig. 77). The effect of age is not clear (Fig. 78). Table 26 shows an example of detailed measurements of the pattern of distribution within a young Pinus densiflora stand. The pattern of distribution of net production for conifer and broadleaved trees is summarized in Fig. 79. There is not much difference between coniferous and broadleaved species in this respect though the distribution into

Table 26. Pattern of distribution of net production into parts of trees in a 15-year-old forest of <u>Pinus densiflora</u>

	Net production kg/ha/yr	Relative distribution % based on	
		Whole tree	Above or below ground
Leaves	4242	27	29
Branches	2728	17	19
Boles	7495	48	52
Above-ground	14465	92	100
Stump	918	6	75
Root	395	2	25
Underground	1313	8	100
Tree layer	15778	100	
Undergrowth	(+)	(+)	

branches is slightly larger in broadleaved species and the distribution into bole is slightly larger in conifers. Approximately one half of the production allocated to the above-ground growth goes to boles, and about one fifth goes to branches, though these values vary greatly among stands.

Table 27. Distribution of production to bole and branches in old <u>Cryptomeria japonica</u> trees (Satoo, unpublished)

	Tree			
	M_1	M_3	K_4	K_5
Age years	212	163	177	174
Diameter breast height cm	107	83	78	95
Height m	33	38	33	35
Production kg/yr				
Bole	35.3	43.2	25.8	19.4
Branches	33.8	11.1	16.1	6.1
Total	69.1	54.3	41.9	25.5
Bole/total	0.511	0.795	0.615	0.761
Branches/total	0.489	0.205	0.385	0.239

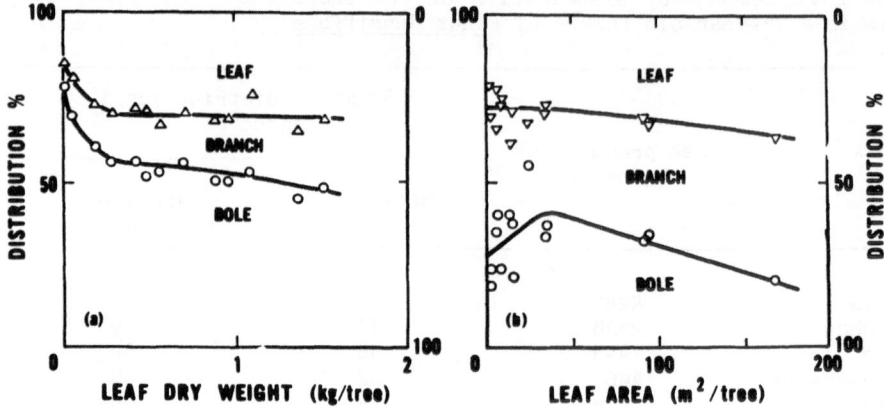

FIGURE 80. The distribution of net above-ground production in relation to dominance of individual trees as expressed as leaf mass for (left) 15-year-old <u>Pinus densiflora</u> and (right) 46-year-old <u>Cinnamomum camphora</u>.

Statistically significant variations in the relative distribution of dry matter between branches and stemwood were found in different genotypes of <u>Pinus virginiana</u> (Matthews <u>et al</u>. 1975) with 32-43 percent of total stem and branch material in stem wood. The pattern of distribution is also very different among individuals within a stand, though this pattern of distribution also varies among stands. Thus in a 15-year-old stand of <u>Pinus densiflora</u> the fractional distribution into bole decreased with increasing tree size (Fig. 80) while in 46-year-old <u>Cinnamomum camphora</u> the distribution into boles was at a maximum in medium sized trees. In a 20-year-old <u>Pinus sylvestris</u> stand medium sized trees again had the highest proportion of growth in boles (Kellomaki 1981) but the smallest trees had the highest fraction of growth in needles in direct contrast to both the <u>P. densiflora</u> and <u>C. camphora</u> examples. The reason for these differences is unclear. Among trees of the same size there are differences in the pattern of distribution. Table 27 shows the pattern of distribution in old trees in natural forests of <u>Cryptomeria japonica</u> in which individual variation in tree shape is very large. Distribution between bole and branches ranged from about 1:1 to 4:1.

4. GROSS PRODUCTION

While many data have been accumulated for net production of forests, information on gross production is very limited. Net production can be measured reasonably well using the harvest method but the estimation of gross production requires the measurement of respiration rates in relation to environmental conditions in order to correct measured net production for respiratory consumption. There are many difficult problems in the measurement of respiration (Negisi 1970, 1977).

Examples of estimates of gross production are given in Table 28. Gross production increases with increases in the product of leaf area index and length of growing season (Fig. 81) and decreases with increasing altitude (Fig. 82). As seen in Fig. 83, between 17 and 30 percent of the difference in gross production by forests of Fagus crenata at altitudes of 550 m and 1550 m, can be attributed to the smaller leaf area index at higher elevations. The remaining difference may be attributed to the lower photosynthetic rate at higher elevation due to reduced radiation as a result of weather conditions (45 percent), to a shorter growing period (45 percent) and to a lower photosynthetic rate due to lower temperatures at higher elevation, (10 percent) (Maruyama 1971). Fig. 84 shows the change in gross production with age of beech forests in Denmark (Möller et al. 1954). Gross production increased rapidly up to the thirtieth year and then gradually decreased. As seen from Table 28, the proportion of net production in gross production in forests ranges from 30 to 70 percent. However, the methods of measurement vary so that comparisons are suspect. The ratio of net production to gross production differs among dominance classes of individual trees (Fig. 85). Thinning removes suppressed trees which have higher rates of respiratory consumption relative to gross production. Net production increases once the recoveries of leaf biomass and gross production are complete (Boysen Jensen 1932). As seen from Fig. 84, consumption by respiration increases with stand age as the result of an increase in non-photosynthetic tissues consequently, in old stands, the rate of net production decreases faster than the rate of gross production.

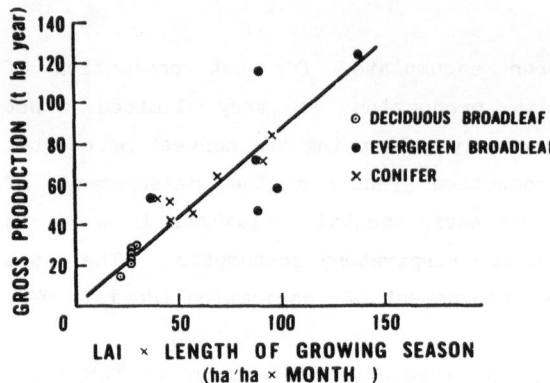

FIGURE 81. The relationship between gross production and the product of leaf area index (LAI) and length of growing season (Kira 1970).

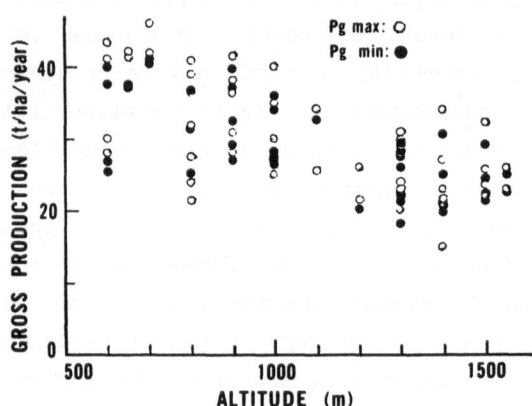

FIGURE 82. Gross production of forests of Fagus crenata in relation to altitude (Marayuma 1970).

Table 28. Examples of estimates of net production (Pn), respiration (r) and gross production (Pg) (t/ha/annum)

Species	Locality	Age	Net production	r	Pg	Pn/Pg	Notes	Source
European beech	Denmark	8	7.5	6.4	13.9	0.54		Moller et al. 1954
European beech	Denmark	25	13.5	8.8	22.3	0.61		Moller et al. 1954
European beech	Denmark	46	13.5	10.0	23.5	0.57		Moller et al. 1954
European beech	Denmark	85	11.3	10.1	21.4	0.53		Moller et al. 1954
European beech	Denmark	22-24	13.6	5.8	19.4	0.70		Boysen Jensen 1932
European ash	Denmark	35	13.5	8.0	21.5	0.63	from figure	Moller 1945
European ash	Denmark	12	6.8	3.1	9.9	0.69	Thinned	Boysen Jensen 1930
European ash	Denmark	18	5.2	3.6	8.8	0.59		Boysen Jensen 1930
European ash	Denmark	12	5.8	2.8	8.5	0.68	Unthinned	Boysen Jensen 1930
European ash	Denmark	18	6.2	2.8	9.0	0.69		Boysen Jensen 1930
European ash	Denmark	22-24	7.4	3.0	10.5	0.71		Boysen Jensen 1932
Fagus crenata	Niigata	30-70	15.3	12.2	27.5	0.56		Maruyama et al. 1968
Cinnamomum camphora	Tiba	46	15.0	32.6	47.6	0.32		Satoo 1968
Laurel forest	Kagosima	-	20.6	52.4	73.1	0.28		Kimura 1960
Tropical rain forest	Cote d'Indies	-	13.4	39.1	52.5	0.26		Muller & Nielsen 1965
Tropical rain forest	Thailand	-	28.6	94.6	123.2	0.23		Kira et al. 1967
Abies veitchii - A. mariesii	Nagano	15	7.4	12.5	19.9	0.37		Kimura et al. 1968
Abies veitchii - A. mariesii	Nagano	mature tree	11.1	28.9	40.0	0.28		Kimura et al. 1968
Pinus densiflora	Tiba	15	15.8	38.1	53.9	0.29		Satoo 1968

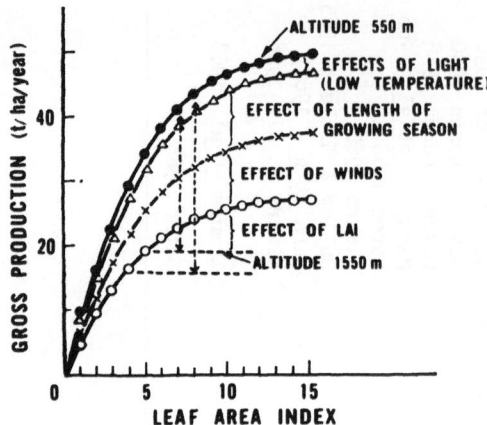

FIGURE 83. Diagrammatic explanation for the decrease in gross production rate with altitude in natural forests of Fagus crenata (Marayuma 1971).

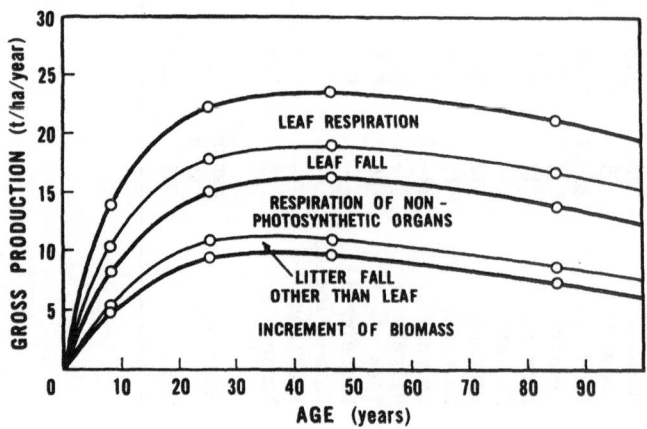

FIGURE 84. The change in gross production and its distribution with age in forests of Fagus sylvatica (Moller et al. 1954).

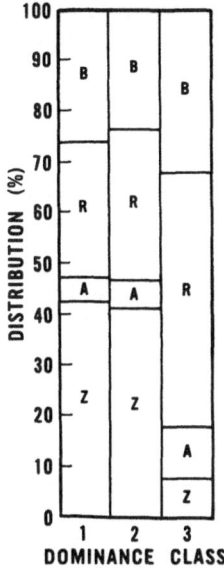

FIGURE 85. The pattern of distribution of gross production by individual tree class in unthinned young·stands of European ash. (B = leaf fall; R = respiration; A = branch fall; Z = increment of biomass) (Boysen and Jensen 1930).

5. ENERGY FIXATION

As far as primary production is concerned, the use of dry weight as the measure of production is very simple and useful. However, when primary production is considered in relation to secondary production and decomposition as a part of the overall flow of material in an ecosystem in which component parts have very different chemical compositions it is more convenient to use energy as the common unit of measurement. The energy content of plants and animals can be measured with a calorimeter, but where direct measurement is not feasible, the energy content of plants may be assumed either from their hypothetical composition (Hellmers and Bonner 1959), or from measurements of similar material. Energy values vary relatively little among tissue types or species within a genus (Table 29). While variation does occur both within and between species (Ovington and Heitkamp 1960; Neenan 1980; Strong 1980) average values for different forest types are quite similar (Table 30). Detailed study suggests energy values may change with time and be affected by the relative dominance of a tree within a stand (Madgwick 1970a) and, for foliage, with position in the crown. However, these effects, while statistically significant, are of negligible absolute magnitude. Stand age does not appear to affect energy values (Madgwick 1970a) and unpublished data for needles of Pinus taeda from 20 plantations in Virginia showed no significant geographic variation or effect of fertilisation.

Table 29. Caloric values (KJ/g) of components of five Pinus species on a dry weight basis (Ovington 1961; Madgwick 1970; Nemeth 1972; Madgwick et al. 1977)

	P. elliottii	P. radiata	P. sylvestris	P. taeda	P. virginiana
Needles - Av.	21.7	-	-	21.4	-
Needles - 1 yr	-	20.3	20.9	-	21.1
Live branches	21.3	19.5	20.5	21.1	20.4
Dead branches	21.9	19.6	20.7	21.3	-
Stem wood	21.0	19.0	-	20.8	19.8
Stem bark	21.5	19.7	-	21.5	19.6
Stem	-	-	20.0	-	-
Male strobili	-	21.1	-	-	21.0
Cones	-	20.0	19.7	-	-
Roots					
5 mm	-	-	20.0	-	-
5 mm	-	-	15.0	-	-

Table 30. Caloric value of tree tissues on a dry weight basis (KJ/g) (Ovington and Heitkamp 1960; Jordan 1971; Neenan 1980; Singh 1980; Madgwick et al. 1981)

	Leaves	Branches	Stem wood	Stem bark	Stem
Temperate Deciduous					
Hardwood	20.6	20.1	20.0	19.4	19.3
Conifer	20.8	-	-	-	20.2
Evergreen					
Hardwood	21.7	17.8	18.6	18.7	16.8
Conifer	20.6	-	-	-	19.8
Tropical	17.3	-	-	-	16.7

Table 31. The efficiency of solar energy use (annual production/visible radiant energy for the growing season) for long term storage as woody tissue and short term storage as leaves, fruit and other litter in forests (Jordan 1971)

Climatic zone	Production term	10^6KJ/m^2/year		Efficiency %	
		Min.	Max.	Min.	Max.
Tropical	Long	5.9	16.0	0.28	1.25
	Short	3.0	36.3	0.15	1.15
Temperate	Long	7.9	46.4	0.57	3.14
	Short	5.4	11.9	0.37	0.84

The efficiency of forests in using solar energy for primary production may be obtained by dividing the primary production expressed as energy content in a given time period by the incident solar energy upon the forest for the same time period. Two values of solar energy namely the total incident energy and the energy effective for photosynthesis (visible light, wavelengths 0.4 to 0.7 um) and two values of primary production namely net production and gross production have been used for such estimates. As a result care must be used in interpreting published data if confusion is to be avoided. Table 31 contains examples of the efficiency of energy utilisation using net production and visible light energy for forests (Jordan 1971). When the duration and timing of the growing period is taken into account, differences among forest types in efficiency are much smaller than their differences in annual production. Jordan (1971) concluded that long term energy storage was similar for both tropical and temperate forests but that short term storage was greater in tropical conditions. Calculations of the efficiency of use of solar energy like this have been made by many workers but reliability of the results is poor as many assumptions must be made in order to complete such calculations. For example, simultaneous measurements of solar energy and production values have rarely been made. Solar energy measurements are frequently obtained from an observatory distant from the site of primary production measurement so that discrepancies in weather conditions between the two locations are possible. Moreover, net primary production measurements on undergrowth and roots have rarely been made and production of these components has been either neglected or estimated on the basis of untested assumptions. There are difficulties in estimating the consumption of gross production by respiration. The amount of work and instrumentation involved in accurately estimating total productivity and efficiency of energy use are such that relatively few data of this type can be made available to cover the wide diversity of forest types and growing conditions.

6. FACTORS AFFECTING RATES OF PRODUCTION

We know that various environmental factors affect photosynthetic efficiency of forest trees and that such relationships are discussed in many textbooks of ecology and physiology. In previous chapters we have described how the dry matter production of forests is affected by various environmental factors and by the management practices of foresters. However dry matter production is not only dependent on rates of photosynthesis; many intermediate processes are involved. Monsi (1960) has described these production processes schematically. Supposing, for simplicity, that the mean photosynthetic rate (a) and mean respiration rate by leaves (r) and non-photosynthetic tissues (r') do not change in a given time period, net production (Pn) may be expressed as:

$$Pn = F(a-r) - C.r' \dots\dots\dots\dots\dots\dots\dots\dots\dots\dots \quad (6.1)$$

where F is the amount of leaves and C the amount of non-photosynthetic tissue (either for individual plants or for plant communities occupying a unit ground area). Using subscripts to indicate values of Pn, F, and C at successive points in time, t_0, t_1, t_2, etc., then at time 0:

$$Pn_0 = F_0 (a-r) - C_0.r' \dots\dots\dots\dots\dots\dots\dots\dots\dots \quad (6.2)$$

Pn_0 is distributed to make additional photosynthetic or non-photo-synthetic tissues and added to F_0 and C_0, thus:

$$Pn_0 = \Delta F_0 + \Delta C_0 \dots\dots\dots\dots\dots\dots\dots\dots\dots\dots \quad (6.3)$$

and $F_1 = F_0 + \Delta F_0$ and $C_1 = C_0 + \Delta C_0$ $\dots\dots\dots\dots\dots\dots$ (6.4)

at time t_1

$$Pn_1 = F_1 (a-r) - C_1.r' \dots\dots\dots\dots\dots\dots\dots\dots\dots \quad (6.5)$$

and so on.

So we see that net production is not only dependent on the rates of photosynthesis and respiration per unit of tissue, but also on the amount of leaf and non-photosynthetic tissue present. Further, when we take time into account the distribution of dry matter production between new leaf and non-photosynthetic tissue and the relative death rates of these two types of tissue we find a variety of factors playing important roles in determining the magnitude of production in subsequent periods. Moreover, the photosynthetic tissues vary among species in their area to weight ratios. Consequently, we might expect the dry matter produced over a growing season to be dependent on the amounts of different tissues present at the beginning of the season, the rates of photosynthesis and respiration, the distribution of new photosynthate to different types, specific leaf area and the death rates of different tissue types. We do not have enough information, particularly concerning photosynthetic and respiration rates, to thoroughly test such a hypothetical model of forest growth over the range of forest types and site fertility that exists. However, experiments with seedlings of sixteen North American hardwood tree species demonstrated that differences in net assimilation rates, distribution patterns of growth and specific leaf area were all important in influencing comparative growth in the first year (Newhouse 1968; Newhouse and Madgwick 1968).

The following analysis and descriptions are highly simplified but help to demonstrate the factors controlling forest productivity and may form the basis for more sophisticated experimentation and data analysis.

1. LEAF EFFICIENCY

The rate of dry matter production (P) can be divided into two factors, the amount of leaves (F) and rate of production per unit amount of leaf (P/F) as

$$P = F \cdot \frac{P}{F} \qquad \dots \dots \dots \dots \dots \dots \dots \dots \dots \dots \dots \dots \dots \dots \quad (6.6)$$

When P is gross production, P/F is the mean rate of gross photosynthesis per unit of leaf. When P is net production, P/F is net rate of photosynthesis obtained per unit of leaf material. The ratio of the amount of non-photosynthetic tissues, C, to the amount of leaf was called the C/F ratio by Iwaki (1959), and from (6.1) it follows that

$$\frac{Pn}{F} = (a - r) - \frac{C}{F} \cdot r' \qquad \dots \dots \dots \dots \dots \dots \dots \dots \dots \dots \dots \dots \quad (6.7)$$

Alternatively, the rate of net production per unit area may be expressed as the product of the rate of photosynthesis per unit of foliage multiplied by the ratio of net production to gross production shown in Table 5.7, i.e.

$$\frac{Pn}{F} = \frac{Pg}{F} \cdot \frac{Pn}{Pg} \quad \dots\dots\dots\dots\dots\dots\dots\dots\dots\dots\dots\dots\dots\dots\dots \quad (6.8)$$

For forests, information on gross production is strictly limited so Pn/F is used to express leaf efficiency. This leaf efficiency is a simplified version of 'unit leaf rate' or 'net assimilation rate' (NAR) calculated in classical growth analysis (Evans 1972). Classical growth analysis has usually been applied to agricultural crops but has been used with forests (Ovington 1957) and tree seedlings (Newhouse and Madgwick 1968; Roberts and Wareing 1975). For forests it is general practice simply to use leaf biomass and net production on an annual basis and measured by the harvest method to determine F and Pn, respectively. In evergreen forests, the total leaf biomass and its age composition change during the year and the values of Pn/F obtained are very approximate. Yoda (1971) called the ratio Pn/F determined this way the 'net production rate' to distinguish it from 'net assimilation rate' as used by classical growth analysis. By the study of Pn/F ratios we can find out whether variations in net production are due to differences in the amount of leaves or caused by the difference in leaf efficiency. Such information may be useful in developing methods for increasing production. However, it must be pointed out that as leaf biomass increases the radiation incident on the leaves of the lower crown decreases and the photosynthetic rate of the canopy as a whole decreases. In other words, leaf efficiency and leaf amount are mutually dependent.

2. AMOUNT AND EFFICIENCY OF LEAVES

The approach outlined above can be used to analyze the production relations of individual trees within a forest stand or variations between forest stands per unit ground area. As shown in Fig. 86, the net above-ground production of individual trees within a 21-year-old stand of Larix leptolepis is linearly proportional to their foliage amounts and the relationship can be expressed by a straight line

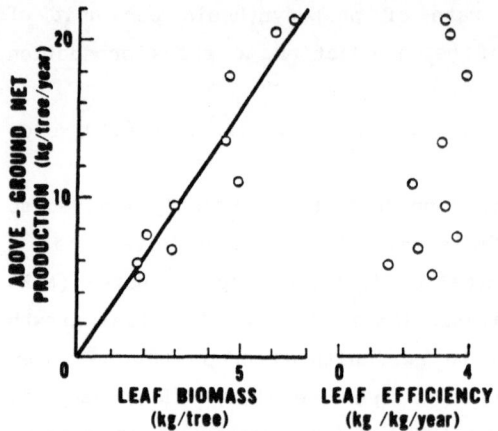

FIGURE 86. The relationship between net annual production by individual trees within a 21-year-old plantation of <u>Larix leptolepis</u> and both leaf biomass and leaf efficiency (Satoo 1971).

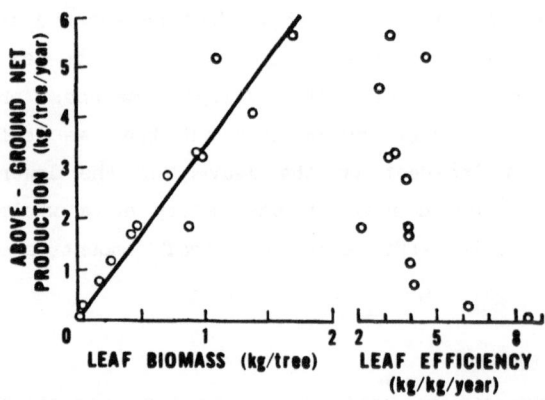

FIGURE 87. The relationship between net annual production by individual trees within a 15-year-old forest of <u>Pinus densiflora</u> and both leaf biomass and leaf efficiency (Satoo 1968).

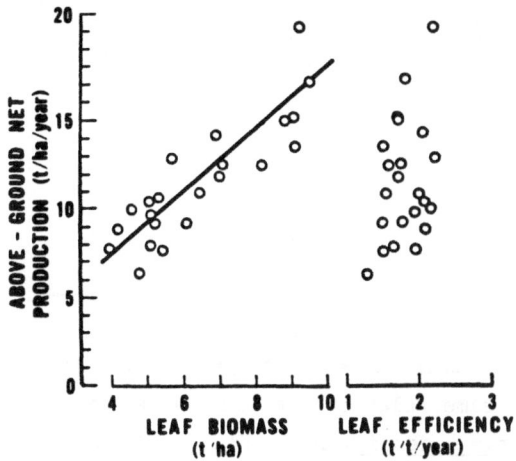

FIGURE 88. The relationship between net above-ground production of forests of <u>Pinus densiflora</u> and their leaf mass and leaf efficiency.

Table 32. Examples of estimates of leaf efficiency using different methods

	Method			Source
	From stand values	From equation 6.6	Mean of sample trees	
g/m²/annum				
Betula maximowicziana	147	148	152	Satoo 1974a
Thujopsis dolabrata	109	111	89	Satoo <u>et al.</u> 1974
Metasequoia glyptostroboides	183	183	182	Satoo 1974d
g/g/annum				
Betula maximowicziana	3.17	3.28	3.74	Satoo 1974a
Thujopsis dolabrata	0.44	0.45	0.37	Satoo <u>et al.</u> 1974
Metasequoia glyptostroboides	3.62	3.62	3.62	Satoo 1974d

Table 32. The foliage mass and efficiency of a forest of Pinus thunbergii on a coastal sand-dune with and without large amounts of added fertilizer (Sakurai, unpublished)

		Unfertilized	Fertilized
Above-ground net productivity	t/ha/yr	8.78	14.20
Leaf mass	t/ha	6.96	7.85
Leaf efficiency	t/t/yr	1.26	1.81
Percentage of new leaf	%	57	59

passing through the origin. In this case, net production is not related to leaf efficiency. The same relationship was found also among trees within stands of Pinus densiflora (Fig. 87), Betula maximowicziana (Satoo 1970b), Abies sachalinensis (Satoo 1974c), Thujopsis dolabrata (Satoo et al. 1974), and Metasequoia glyptostroboides (Satoo 1974d). In some stands a slight correlation seemed to exist between net production and leaf efficiency among suppressed trees, but these correlations were not significant. The same relationship was found among forest stands as shown in Fig. 88. Net above-ground production per unit ground area of forest stands was linearly proportional to leaf biomass and unrelated to leaf efficiency, though the variation is rather large. The same relation was found among stands of Betula maximowicziana (Satoo 1970) and Larix leptolepis.

Leaf efficiency for a forest stand can be approximated by three different methods. The straight lines on the left hand graphs of Figs. 86 and 87 represent equation (6.6) and the slope of these lines represents leaf efficiency. The slope can be easily determined by means of least squares regression. The second method is to calculate the mean value of P/F for sample trees or stands. The third method is to divide the value of net above-ground production estimated for the stand by the leaf biomass of the stand. The differences among the values obtained by the different methods is small, as seen from Table 32.

3. EXAMPLES OF ANALYSIS OF NET PRODUCTION

As seen in Fig. 88, there is a wide range in net above-ground production among stands having the same leaf biomass and in leaf efficiency among stands having the same above-ground net production when we combine all the data available and, therefore, include a wide variety of forest stands. From these data, some subsets showing rather large differences in net above-ground production may be selected for closer study.

Fig. 89 contains data for the stands shown in Fig. 70 and in which above-ground net production increased with relative stand density. Net production increased with leaf biomass and was inversely related to leaf efficiency. This result would be predicted from the hypothetical curvilinear relationship between growth and foliar biomass (Watson 1952; Waring et al. 1981). Fig. 90 includes the same stands as shown in Fig. 72 which showed that net above-ground production changed with age. Net production was independent of leaf biomass but increased with increasing leaf efficiency. The lowest leaf efficiencies occurred in the oldest stands where C/F would be high. Fig. 91 contains the same stands as shown in Fig. 68 which demonstrated that above-ground net production increased with increasing site index when the latter was expressed as relative height. Net production increased with both amount and efficiency of leaves. A similar trend was found in a heavily fertilized coastal sand dune plantation of Pinus thunbergii. Both the amount and efficiency of leaves were larger in fertilized plots which produced larger amounts of dry matter (Table 33). As seen from these four examples, differences in net production were due, in one case, to differences in leaf biomass, in another to leaf efficiency, and in two cases to a combination of the amounts and efficiencies of leaves. There could be other combinations of causal elements. The first example suggests that higher production could be obtained by increasing leaf biomass either as a result of keeping a higher relative stand density or by other means such as fertilization. The second example indicates that leaf efficiency changes with age. The third and fourth examples suggest processes by which soil fertility affects production and tree growth.

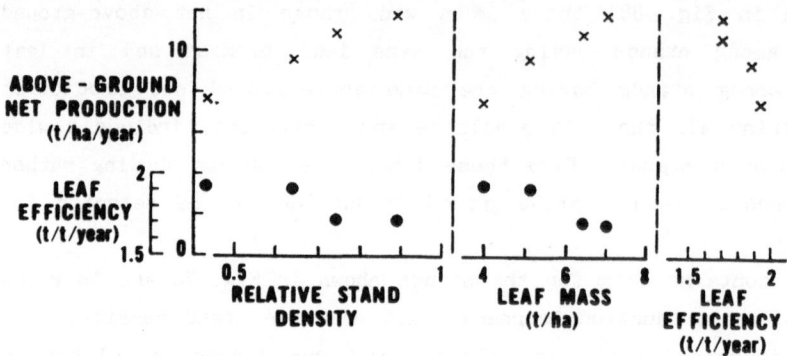

FIGURE 89. Factors affecting the above-ground net production of 43-to 46-year-old forests of _Pinus densiflora_ of different stand density (Hatiya and Kobayashi 1965).

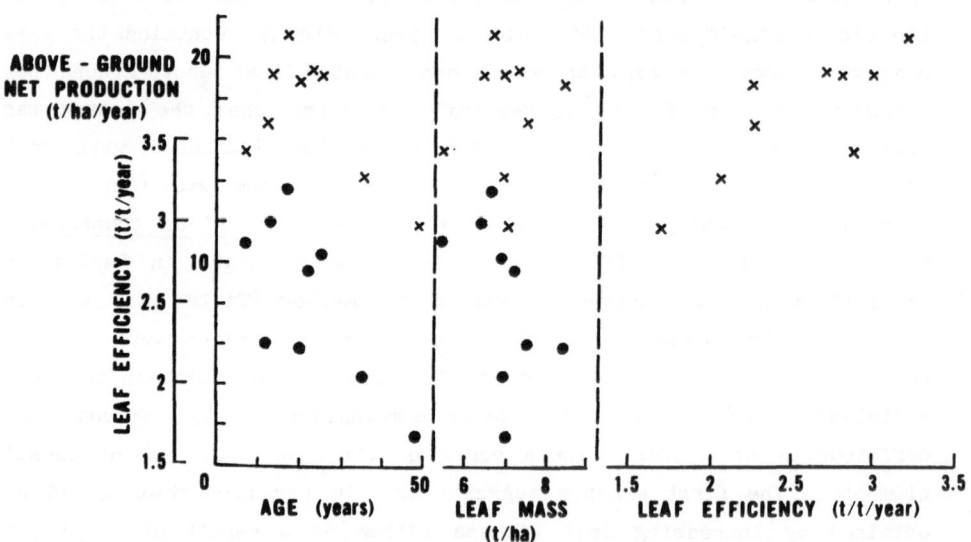

FIGURE 90. Factors affecting the above-ground net production of _Pinus densiflora_ forests of different ages (Hatiya and Tochiaki 1968).

Table 33. Efficiency of leaves of overstorey trees and undergrowth

Overstorey canopy	LAI	Overstorey	Undergrowth	Source
Larix leptolepis	4.3	2.95	1.04	Satoo 1973b
Metasequoia glyptostroboides	8.5	1.90	1.04	Satoo 1973d
Cinnamomum camphora	4.9	3.38	0.89	Satoo 1968
Betula ermanii	2.9	1.92	0.52	Tadaki and Hatiya 1970
Betula ermanii	4.2	1.89	0.64	Tadaki and Hatiya 1970

The final example is a group of plots with very similar amounts of above-ground net production (Fig. 92). The small sample plots were located in young natural regeneration of Pinus densiflora and were felled amd measured in November 1967, March, April, October and November 1968. The slight differences in net production cannot be related to either the amount, or the efficiency, of leaves taken independently but leaf efficiency decreased with increasing leaf biomass and the two factors compensated for each other to result in minimal differences in net production. Leaf efficiency increased with the amount of new leaves as a proportion of total leaf mass which accords with the fact that photosynthetic rate decreases with leaf age. It has also been observed that leaf efficiency of plantations of Picea abies in Japan was higher in stands which had a larger proportion of new needles to total needle biomass. Leaf efficiency seems to decrease with stand age and is higher on better sites or in fertilized forests. There is little information on the leaf efficiency of the understorey in forests but the data in Table 33 confirm that leaf efficiency of undergrowth is lower than that of the overstorey tree layer. The lower efficiency of understorey plants presumably reflects the low levels of radiant energy they receive.

4. FOREST TYPE

Leaf efficiencies have been calculated for a range of plots for which both net above-ground production and leaf biomass are known. The results are arranged in order of efficiency by forest type in Fig. 93.

128

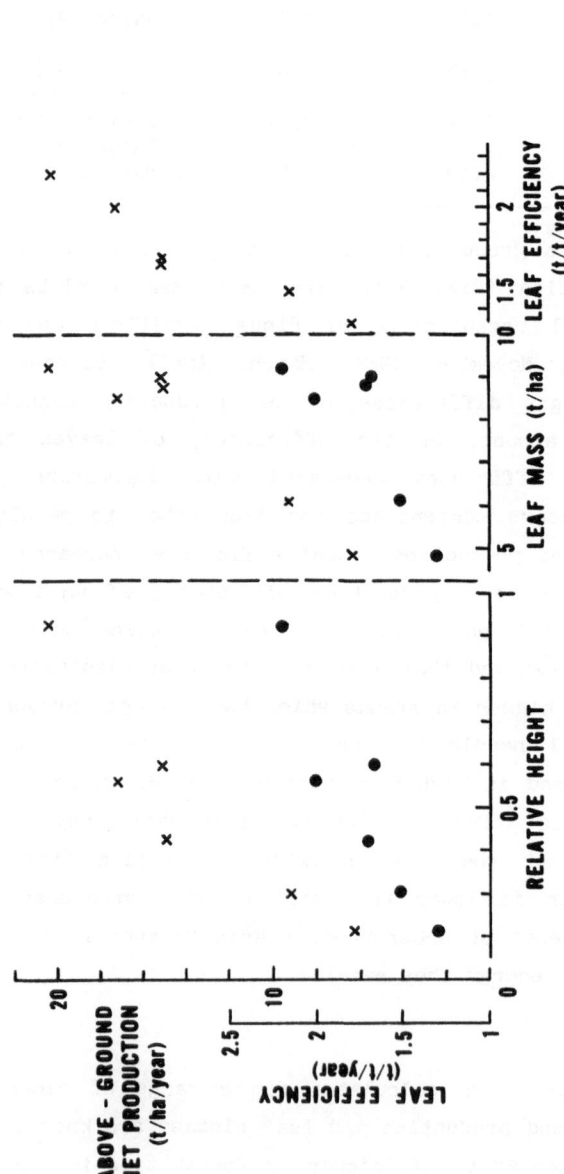

FIGURE 91. Factors affecting the above-ground net production of 20- to 21-year-old forests of Pinus densiflora of different site index (Hatiya et al. 1966).

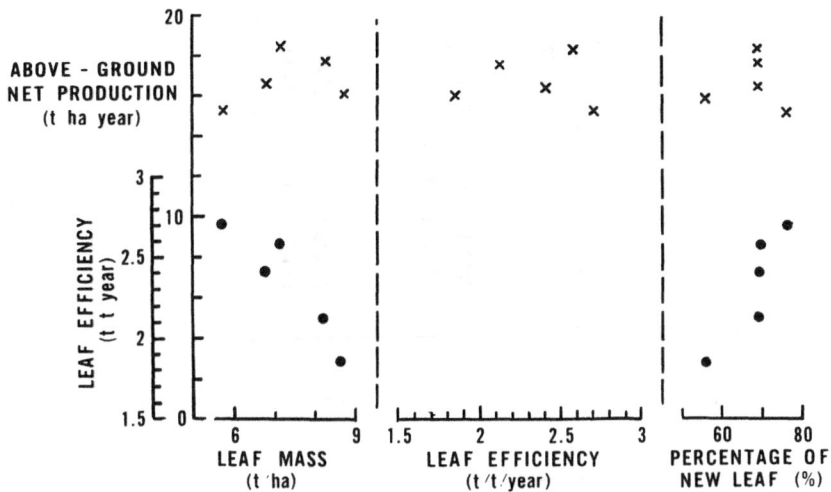

FIGURE 92. Factors affecting the above-ground net production in a group of 20-year-old forests of <u>Pinus densiflora</u> (Satoo 1970).

Among coniferous forests, deciduous species like larch have the highest efficiency, pines are intermediate and other evergreen conifers lowest. Differences between evergreen and deciduous broadleaved forests were very slight with values very close to those for deciduous conifers. The order of leaf efficiency seems approximately inversely correlated with leaf biomass. The mean values of leaf efficiency of coniferous forests are plotted against their mean values of leaf biomass in Fig. 94. Leaf efficiency decreased with increasing leaf biomass and these two parameters were linearly related when plotted on a double logarithmic scale. The slope of the regression is close to -1.0, which implies that leaf biomass and leaf efficiency are inversely proportional and, on average, production per hectare is constant irrespective of leaf biomass. A similar linear relationship was reported for crop plants on a linear scale by Watson (1958) and for forests (including both coniferous and broadleaved) on a logarithmic scale by Yoda (1971). The decrease in leaf efficiency in forests with an increase in leaf biomass could be explained by a combination of lower mean photosynthetic rate of their canopies due to reduced light intensity in the lower canopy and a lower mean photosynthetic rate due to an increase in average leaf age (Fig. 48).

130

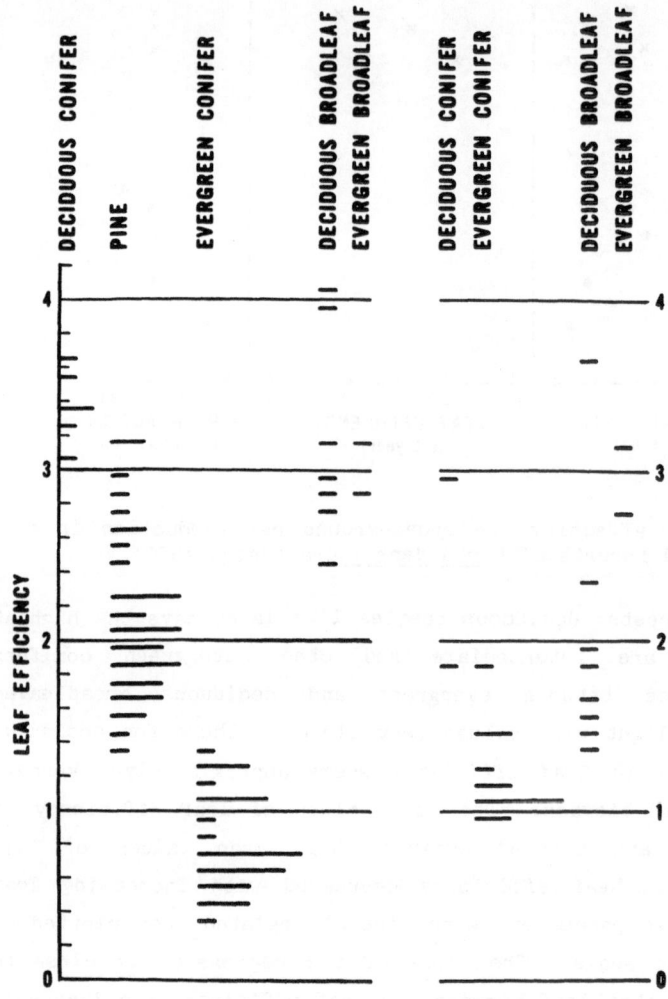

FIGURE 93. Leaf efficiency as affected by forest type. Left, on a leaf dry weight basis(t/t/annum); right, on a leaf area basis (t/ha/annum).

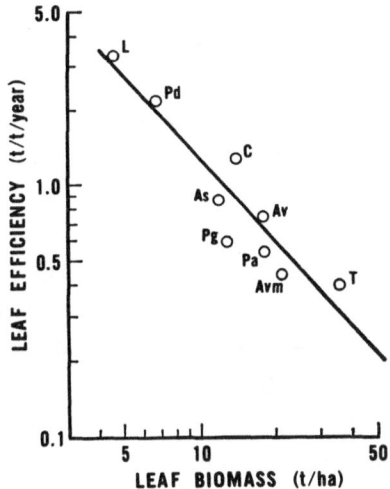

FIGURE 94. The relationship between leaf efficiency and leaf biomass in forests of different conifer species (Satoo 1971). (As, <u>Abies sachalinensis</u> (3 plots); Av, <u>Abies veitchii</u> (7); Avm, mixed <u>A. veitchii</u> and <u>Abies mariesii</u> (1); L, <u>Larix leptolepis</u> (4); Pg, <u>Picea glehnii</u> (1); Pd, <u>Pinus densiflora</u> (37); C, <u>Chamaecyparis obtusa</u> (1); T, <u>Thujopsis dolabrata</u> (3); Pa, <u>Picea abies</u> (5).

5. THE EFFICIENCY OF LEAVES IN THE PRODUCTION OF BOLE WOOD

Bole wood production per unit amount of leaf has been used as the measure of leaf efficiency in forestry for a long time. According to Adams (1928), Hartig used this concept in 1891. Burger (1929-1953) published many studies on this way of evaluating leaf efficiency which has since been used by many researchers. In his book on forest ecology, Baker (1950) devoted several pages to this approach. However, the 'efficiency of leaves in the production of bole wood' is a complex variable. The 'efficiency of leaf in the production of bole wood' (Ps/F) is the product of leaf efficiency in terms of net production (Pn/F) times the fraction of dry matter production distributed to bole wood (Ps/Pn), i.e.

$$\frac{Ps}{F} = \frac{Pn}{F} \cdot \frac{Ps}{Pn} \quad \dots\dots\dots\dots\dots\dots\dots\dots\dots\dots\dots\dots\dots\dots\dots \quad (6.9)$$

This equation may be expressed in terms of gross photosynthesis by combining equations 6.8 and 6.9 to obtain the form

$$\frac{Ps}{F} = \frac{Pg}{F} \cdot \frac{Pn}{Pg} \cdot \frac{Ps}{Pn} \quad \dots\dots\dots\dots\dots\dots\dots\dots\dots\dots\dots\dots\dots\dots \quad (6.10)$$

Net production of bole wood and leaf biomass may be estimated fairly exactly using the harvest method, but the estimation of gross production involves several difficulties. This approach may be applied both to individual trees within a stand and to stands as a whole. Fig. 95 shows the relationships between the 'efficiency of leaves to produce bole wood' and both leaf efficiency and the relative distribution of photosynthate by individual trees in a stand of Pinus densiflora. Fig. 96 shows the comparable relationship for stands of Pinus densiflora including many different types of forest. In both cases, 'efficiency to produce bole wood' increased with both leaf efficiency and the relative amount of photosynthate distributed to bole.

Crop ecologists have concluded that, in a variety of agricultural crops, the trend towards higher yields per hectare have largely resulted in a shift in the distribution of dry matter production into the 'crop' part of the plant (Gifford and Evans 1981). In several grain crops biological yield (total dry matter production) has not risen. Moreover Gifford and Evans suggested that there is an upper limit, not yet reached, beyond which any further increase in the fraction of production going to the crop would be counterproductive. The maximum photosynthetic rate per unit of leaf area does not appear to have been increased by genetic selection. These conclusions have important implications for future forest biomass studies. We have indicated that a variety of factors interact to control forest productivity. These are the rates of photosynthesis and respiration, annual foliage production, leaf longevity, and the distribution of photosynthate within the plant. Net photosynthesis in trees is known to be under genetic control (Roberts and Wareing 1975) but has not been shown to have any major effect on overall production per se. For canopies as a whole photosynthetic efficiency decreases with foliage mass and may, under some conditions, be above the optimum for maximum productivity (Waring et al. 1981). The inverse relationship between foliage mass and foliage efficiency (Fig. 94) raises interesting questions concerning optimum needle longevity. Certainly, the evergreen habit has been suggested as a reason for the relatively high levels of forest productivity (Ovington 1957) since the foliage mass is potentially available to photosynthesize whenever weather conditions

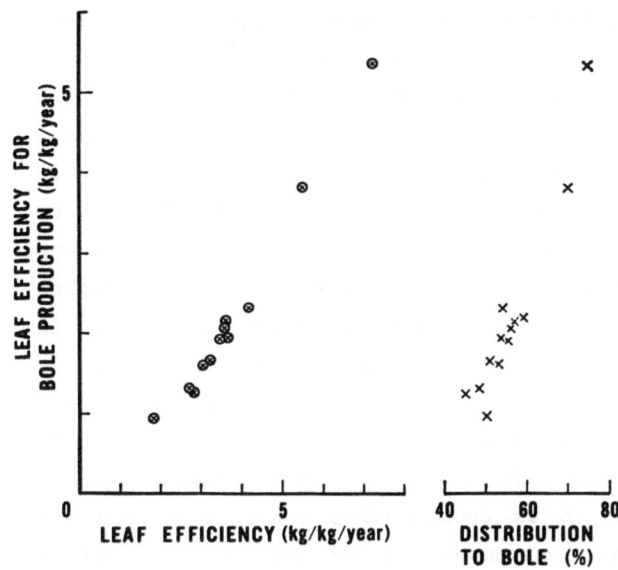

FIGURE 95. The 'efficiency of leaves in the production of boles' as related to both the efficiency of leaves to produce above-ground net production and the percentage of net production devoted to boles in individual trees in a 15-year-old forest of <u>Pinus densiflora</u>.

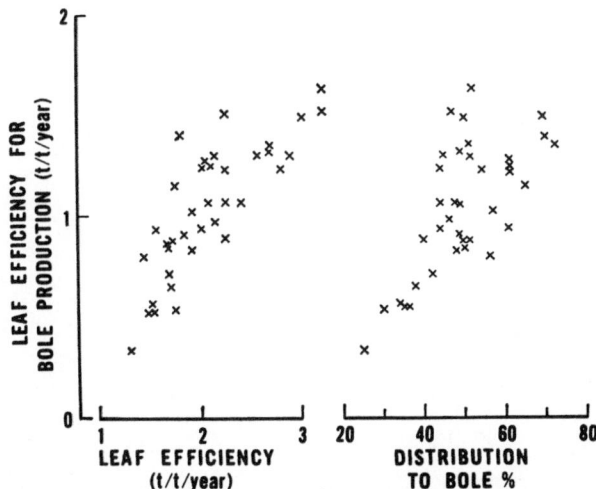

FIGURE 96. The 'efficiency of leaves in the production of boles' as related to both the efficiency of leaves to produce above-ground net. production and the percentage of net production devoted to boles in forest stands of <u>Pinus densiflora</u>.

are suitable. But is there an optimum needle longevity for forest productivity? The distribution of photosynthate among tree components is also known to be under genetic control (Matthews et al. 1975), but compared with agricultural crops, trees naturally have a high fraction of above-ground production in stemwood (Fig. 95, 96). There must be some optimum pattern of distribution for maximum yield. It can be hypothesized that the optimum pattern of distribution should involve a change with stand condition from one of a high proportion in foliage during canopy closure to a higher proportion in stems in later life which is the pattern found in Pinus sylvestris by Ovington (1957). Our ability to understand the mechanisms affecting dry matter production of forests and to use this knowledge to improve yields of harvestable products will depend on the quality and comprehensiveness of future studies. As the questions become clearer it is obvious that there is a need for studies to go beyond the determination of the standing crop and to include estimates of those parameters known to be associatd with productivity.

REFERENCES

Adams, W.R. 1928: Effect of spacing in Jack Pine plantation. Vt. Agr. Expt. Sta. Bull. 282. pp. 51.

Alemdeg, I.S. 1980: Manual of data collection and processing for the development of forest biomass relationships. Canadian For. Serv. Info. Rept P1-X-4, 38 p.

Ando, T., Hatiya, K., Doi, K., Kataoka, H., Kato, Y. and Sagaguchi, K. 1968: Studies on the system of density control of sugi (Cryptomeria japonica) stands. Bull. Govt. For. Expt. Sta., Tokyo. 209: 1-76*.

Ando, T., and Takuchi, I. 1968: (Change of biomass in young plantation of Cryptomeria japonica from April through November). Abstracts, 79th Mtg., Jap. For. Soc. (1968): 39-40**.

Assmann, E. 1961: Waldertragskunde. BLV Verlagsges., Muchen 472 p. (Engl. transl.: The principles of forest yield study. (Gradiner, S.H., transl., Davis, P.W., ed.) Pergamon Press, Oxford 1970. pp. xiv + 506.

Attiwill, P.M. 1962: Estimating branch dry weight and leaf area from measurements of branch girth in Eucalyptus. For. Sci. 8: 132-41.

Attiwill, P.M. 1966: A method for estimating crown weight in Eucalyptus, and some implications of relationships between crown weight and stem diameter. Ecology 47: 795-804.

Attiwill, P.M. and Ovington, J.D. 1968: Determination of forest biomass. For. Sci. 14:13-5.

Axelsson, B., Gärdefors, D., Hyttebron, H., Lohm, U., Persson, T. and Tenow, O. 1972: Estimation of leaf number and leaf biomass of hazel Corylus avellana by two methods. Oikos 23: 281-3.

Baker, F.S. 1950: Principles of Silviculture. McGraw-Hill, New York U.S.A. pp. 414.

Bandola-Ciolczyk, E. 1974: Production of tree leaves and energy flow through the litter in Tilio - Carpinetum association (International Biological Programme area). Zaklad Ochrony Przyrody, Studia Naturae No. 9; 29-91.

Barney, R.J., Van Cleve, K. and Schlentner, R. 1978: Biomass distribution and crown characteristics in two Alaskan Picea mariana ecosystems. Can. J. For. Res. 8: 36-41.

Baskerville, G.L. 1965: Estimation of dry weight of tree components and total standing crop in conifer stands. Ecology 46: 867-9.

Baskerville, G.L. 1966: Dry-matter production in immature balsam fir stands: Roots, lesser vegetation, and total stand. For. Sci. 12: 49-53.

Baskerville, G.L. 1972: Use of logarithmic regressions in the estimation of plant biomass. Can. J. For. Res. 2: 49-53.

Beauchamp, J.J. and Olson, J.S. 1973: Corrections for bias in regression estimates after logarithmic transformation. Ecology, 54: 1403-7.

Boysen Jensen, P. 1910: Studier over skovtraernes forhold til lyset. Tidsskr. f. Skovvaesen 22: 1-116.

Boysen Jensen, P. 1930: Undersogelser over stoffprøduktionen i yngre bevoksninger af ask og bog. 2. Forstl. Forsogsv. Danmark 10: 365-391.

Boysen Jensen, P. 1932: Die Stoffproduktion der Pflanzen. Gustav Fischer, Jena., 108 p.

Boysen Jensen, P., and Müller, D. 1927: Undersogelser over stoffproduktion i yngre bevoksninger af ask og bog. Forstl. Forsogsv. Danmark 9: 221-268.

Bradstock, R. 1981: Biomass in an age series of Eucalyptus grandis plantations. Aust. For. Res. 11: 111-27.

Bradu, D. and Mundlak, Y. 1970: Estimation in lognormal linear models. J. Amer. Stat. Assoc. 65: 198-211.

Bray, J.R. 1961: Measurement of leaf utilisation as an index of minimum level of primary consumption. Oikos, 12: 70-4.

Bray, J.R. 1962: Root production and the estimation of net productivity. Can. J. Bot. 41: 65-72.

Bray, J.R. 1964: Primary consumption in three forest canopies. Ecology 45: 165-7.

Bray, J.R. and Gorham, E. 1964: Litter production in forests of the world. Adv. Ecol. Res. 2: 101-57.

Brix, H. 1981: Effects of thinning and nitrogen fertilization on branch and foliage production in Douglas fir. Can. J. For. Res. 11: 502-11.

Brown, J.K. 1976: Estimating shrub biomass from basal stem diameters. Can. J. For. Res. 6: 153-8.

Bunce, R.G.H. 1968: Biomass and production of trees in a mixed deciduous woodland. 1. Girth and height as parameters for the estimation of tree dry weight. J. Ecol. 56: 759-75.

Burger, H. 1929: Holz, Blattmenge und Zuwachs. 1. Weymouthfohre. Mitteil. Schweiz. Centralanst. Forstl. Versuchsw. 15: 243-92.

Burger, H. 1935: Holz, Blattmenge und Zuwachs. 2. Die Douglasie. Mitteil. Schweiz. Anst. Forstl. Versuchsw. 19: 21-72.

Burger, H. 1937: Holz, Blattmenge und Zuwachs. 3. Nadelmenge und Zuwachs bei Fohren und Fichten verschiedener Herkunft. Mitteil. Schweiz. Anst. Forstl. Versuchsw. 20: 101-14.

Burger, H. 1940: Holz, Blattmenge und Zuwachs. 4. Ein 80jähriger Buchenbestand. Mitteil. Schweiz. Anst. Forstl. Versuchsw. 21: 307-48.

Burger, H. 1941: Holz, Blattmenge und Zuwachs. 5. Fichten und Föhren verschiedener Herkunft auf verschidenen Kulturorten. Mitteil. Schweiz. Anst. Forstl. Versuchsw. 22: 10-62.

Burger, H. 1942: Holz, Blattmenge und Zuwachs. 6. Ein Plenterwald mittlerer Standortsgüte. Mitteil. Schweiz. Anst. Forstl. Versuchsw. 22: 377-445.

Burger, H. 1945: Holz, Blattmenge und Zuwachs. 7. Die Lärche. Mitteil. Schweiz. Anst. Forstl. Versuchsw. 24: 7-103.

Burger, H. 1947: Holz, Blattmenge und Zuwachs. 8. Die Eiche. Mitteil. Schweiz. Anst. Forstl. Versuchsw. 25: 211-79.

Burger, H. 1948: Holz, Blattmenge und Zuwachs. 9. Die Fohre. Mitteil. Schweiz. Anst. Forstl. Versuchsw. 25: 435-93.

Burger, H. 1950: Holz, Blattmenge und Zuwachs. 10. Die Buche.
  Mitteil. Schweiz. Anst. Forstl. Versuchsw. 26: 419-68.
Burger, H. 1951: Holz, Blattmenge und Zuwachs. 11. Die Tanne.
  Mitteil. Schweiz. Anst. Forstl. Versuchsw. 27: 247-86.
Burger, H. 1952: Holz, Blattmenge und Zuwachs. 12. Fichten im
  Plenterwald. Mitteil. Schweiz. Anst. Forttl. Versuchsw. 28: 109-56.
Burger, H. 1953: Holz, Blattmenge und Zuwachs. 13. Fichten im
  gleichalterigen Hochwald. Mitteil. Schweiz. Anst. Forstl.
  Versuchsw. 29: 38-130.
Conkle, M.T. 1963: The determination of experimental plot size
  and shape in loblolly and slash pines. N.C. State, School of
  Forestry, Tech. Rept. No. 17, 51 p.
Crow, T.R. and Laidly, P.R. 1980: Alternative models for estimating
  woody plant biomass. Can. J. For. Res. 10: 367-70.
Cummings, W.H. 1941: A method for sampling the foliage of a silver
  maple tree. J. For. 39: 382-4.
Cunia, T. 1979a: On tree biomass tables and regression: Some
  statistical comments. In: Frayer, W.E. (ed.) Forest Resource
  Inventories, Colorado State Univ., Fort Collins, Colo., U.S.A.,
  629-642.
Cunia, T. 1979b: On sampling trees for biomass table construction:
  Some statistical comments. In: Frayer, W.E. (ed.) Forest Resource
  Inventories, Colorado State Univ., Fort Collins, Colo., U.S.A.,
  643-664.
Curtin, D.T., Brooks, R.T. and Rennie, J.C. 1980: Testing biomass
  prediction equations with conventional cruise and whole tree
  harvesting techniques. Tennessee Valley Authority, Div. Land For.
  Res., Tech. Note B41, 27 p.
Day, F.P. and Monk, C.D. 1977: Net primary production and phenology
  on a southern Appalachian watershed. Amer. J. Bot. 64: 1117-25.
Droste zu Hulshoff, B.V. 1970: Struktur, Biomasse und Zuwachs eines
  älteren Fichtenbestandes. Forstw. Cbl. 89: 162-171.
Ebermayer, E. 1876: Die gesamte Lehre der Waldstreu mit Rucksicht
  auf die chemische Statik des Waldbaues. J. Springer, Berlin. SS 300 +
  116.
Egunjobi, J.K. 1968: An evaluation of five methods for estimating
  biomass of an even-aged plantation of Pinus caribaea L. Oecol. Plant.
  11: 109-16.
Evans, G.C. 1972: The quantitative analysis of plant growth.
  Blackwell, Oxford, 734 p.
Ewers, F.W. and Schmid, R. 1981: Longevity of needle fascicles of
  Pinus longaeva (Bristlecone pine) and other North American pines.
  Oecologia 51: 107-15.
Finney, D.J. 1941: On the distribution of a variate whose logarithm
  is normally distributed. J. Roy. Stat. Soc., Ser B, 7: 155-61.
Flewelling, J.W. and Pienaar, L.V. 1981: Multiplicative regression
  with lognormal errors. For. Sci. 27: 281-9.
Forward, D.F. and Nolan, N.J. 1961: Growth and morphogenesis in the
  Canadian forest species. IV. Radial growth in branches and main
  axis of Pinus resinosa. Ait under conditions of open growth,
  suppression and release. Can. J. Botany 39: 385-409.
Forrest, W.G. 1968: The estimation of oven dry weight. Aust. For.
  Res. 3: 41-6.

138

Forrest, W.G. and Ovington, J.D. 1970: Organic matter changes in an
    age series of Pinus radiata plantations. J. appl. Ecol. 7: 177-86.
Freese, F. 1961: Relation of plot size to variability: An
    approximation. J. For. 59: 679.
Fujimori, T., Kawanabe, S., Saito, H., Grier, C.C. and Shidei,
    T. 1976: Biomass and primary production in forests of three major
    vegetation zones of the northwestern United States. J. Jap. For.
    Soc. 58: 360-73.
Garcia, O.V. 1974: Ecuaciones altura-diametro para pino insigne.
    Inst. For. Chile, Tech. Note No. 19, 6 p.
Gifford, R.M. and Evans, L.T. 1981: Photosynthesis, carbon
    partitioning and yield. Ann. Rev. Plant Physiol. 32: 485-509.
Gordon, A.G. 1975: Productivity and nutrient cycling by site in
    spruce forest ecosystems. In: Cameron, T.W.M. and Billingsley, L.W.
    (eds.) Energy Flow - Its Biological Dimensions. A summary of the IBP
    in Canada. Royal Society of Canada, Ottawa, 119-126.
Greenland, D.J. and Kowal, J.M.L. 1960: Nutrient content of the
    moist tropical forest of Ghana. Plant and Soil 12: 154-74.
Grieg-Smith, P. 1964: Quantitative Plant Ecology. Butterworths,
    London, xii + 256 p.
Grier, C.C., Vogt, K.A., Keyes, M.R. and Edmonds, R.L. 1981:
    Biomass distribution and above- and below-ground production in young
    and mature Abies amabilis zone ecosystems of the Washington
    Cascades. Can. J. For. Res. 11: 155-67.
Grier, C.C. and Waring, R.H. 1974: Conifer foliage mass related to
    sapwood area. For. Sci. 20: 205-6.
Hagihara, A., Suzuki, M. and Hozumi, K. 1978: Seasonal fluctuations of
    litter fall in a Chamaecyparis obtusa plantation. J. Jap. For. Soc.
    60: 397-404.
Harris, W.F., Kinerson, R.S. and Edwards, N.T. 1977: Comparison
    of belowground biomass of natural deciduous forests and loblolly pine
    plantations. In: Marshall, J.K. (ed.) The Belowground Ecosystem: A
    Synthesis of Plant-Associated Processes. Range Sci. Dept., Sci. Ser.
    No. 26, Colorado State University, Fort Collins, Colo., U.S.A.:
    29-37.
Hatiya, K., and Ando, T. 1965: (Stand density and growth of
    plantations of Cryptomeria japonica). Trans. 75th Mtg., Jap. For.
    Soc. (1964). 340-2**.
Hatiya, K., Doi, K. and Kobayashi, R. 1965: Analysis of growth in
    Japanese red pine (Pinus densiflora) stands. Bull. Govt. For. Expt.
    Sta., Tokyo. 176: 75-88*.
Hatiya, K., Fujimori, T., Tochiaki, K. and Ando, T. 1966: Studies
    on the seasonal variations of leaf and leaf-fall amount in Japanese
    red pine (Pinus densiflora) stands. Bull. Govt. For. Expt. Sta.,
    Tokyo. 191: 101-13*.
Hatiya, K., and Tochiaki, K. 1968: (Analysis of growth in stands of
    Pinus densiflora of high density: relationships between growth and
    age). Abstract, 79th Mtg., Jap. For. Soc., (1968), 31-2**.
Hatiya, K., Tochiaki, K., Narita, T. 1966: (Analysis of growth in
    natural forests of Pinus densiflora: relationships between site
    quality and growth). Trans. 76th Mtg., Jap. For. Soc. (1965):
    161-2**.

Heilman, P.E. and Gessel, S.P. 1963: The effect of nitrogen
     fertilization on the concentration and weight of nitrogen,
     phosphorus, and potassium in Douglas-fir trees. Soil Sci. Soc. Amer.
     Proc. 27: 102-5.
Hellmers, H., and Bonner, J. 1960: Photosynthetic limits of forest
     tree yields. Proc. Am. Soc. For. 1959: 32-5.
Hermann, R.K. 1977: Growth and production of tree roots: A review.
     In: Marshall, J.K. (ed.) The Belowground Ecosystem: A Synthesis of
     Plant-Associated Processes. Range Sci. Dept. Sci. Ser. No. 26,
     Colorado State University, Fort Collins, Colo., U.S.A.: 7-28.
Hirai, S. 1947: Studies on the weight-growth of forest trees. 1.
     Larix leptolepis Grodon of Fuji University Forest. Bull. Tokyo Univ.
     For. 35: 91-105*.
Hitchcock, H.C. and McDonnell, J.P. 1979: Biomass measurement: A
     synthesis of the literature. In Frayer, W.E. (ed.). Forest Resource
     Inventories, Colorado State Univ., Fort Collins, Colo., U.S.A.
     544-595.
Honer, T.G. 1971: Weight relationships in open- and forest-grown
     balsam fir trees. In. Young, H.E. (ed.). Forest Biomass Studies,
     Univ. Maine Press, Orono, Me., U.S.A. 63-78.
Hughes, M.K. 1971: Tree biocontent, net production and litter fall
     in a deciduous woodland. Oikos 22: 62-73.
Iwaki, H. 1959: Ecological studies of interspecific competition in
     plant community. 1. An analysis of growth of competing plants in
     mixed stands of buckwheat and green gram. Jap. J. Bot. 17: 120-38.
Jackson, D.S. and Chittenden, J. 1981: Estimation of dry matter in
     Pinus radiata root systems. 1. Individual trees. N.Z. J. For. Sci.
     11: 164-82.
Jordan, C.F. 1971: Productivity of tropical forest and its relation
     to a world pattern of energy storage. J. Ecol. 59: 127-42.
Kakubari, Y., Maruyama, K. and Yamada, M. 1970: Ecological studies
     on natural beech forest 21. A comparative study of biomass and
     productivity according to altitudes on beech forest. Niigata Norin
     Kenkyu 22: 43-55**.
Kato, R. 1968: (Primary production of plantations of Pinus
     densiflora). In. Satoo, T. (ed.). Interim Rept for 1967, Special
     Project Research "Primary production of managed forests": 30-3***.
Kawahara, T., Iwatsubo, G., Nishimura, T. and Tsutsumi, T. 1968:
     Movement of nutrients in a model stand of Camptotheca acuminata
     Decne. J. Jap. For. Soc. 50: 125-34*.
Kawahara, T., Kanazawa, Y. and Sakurai, S, 1981: Biomass and net
     production of man-made forests in the Philippines. J. Jap. For. Soc.
     63: 320-7.
Kellomaki, S. 1981: Effect of the within-stand light conditions on
     the share of stem, branch and needle growth in a twenty-year-old
     Scots pine stand. Silva fenn. 15: 130-9.
Keyes, M.R. and Grier, C.C. 1981: Above- and below-ground net
     production in 40-year-old Douglas fir stands on low and high
     productivity sites. Can. J. For. Res. 11: 599-605.
Kiil, A.D. 1969: Estimating fuel weights of black spruce and
     alpine fir crowns in Alberta. Bimonthly Res. Notes 25: 31-2.

140

Kimmins, J.P. and Krumlik, G.J. 1973: Comparison of biomass
    distribution and tree form of old virgin forests at medium and high
    elevations in the mountains of south coastal British Columbia,
    Canada. In: Young, H.E. (ed.) IUFRO Biomass Studies. Univ. Maine
    Press, Orono, Me. U.S.A. 315-35.
Kimura, M. 1960: Primary production of the warm temperate laurel
    forest in the southern part of Osumi Peninsula, Kyushu, Japan. Misc.
    Rep. Res. Inst. Natur. Resources 52/53: 36-47.
Kimura, M., Mototani, I. and Hogetsu, K. 1968: Ecological and
    physiological studies on the vegetation of Mt Shimagare. 6. Growth
    and dry matter production of young Abies stand. Bot. Mag. Tokyo 81:
    287-96.
Kira, T. 1970: (Primary productivity and efficiency of energy
    utilisation). In: Interim Rept for 1969, Special Project Research
    "Comparative study of primary productivity of natural forests".
    (Shidei, T., ed.): 85-92***.
Kira, T., and Shidei, T. 1967: Primary production and turnover of
    organic matter in different forest ecosystems of the western
    Pacific. Jap. J. Ecol. 17: 70-87.
Kira, T., et al. 1960: (Growth and Productivity). In: Studies on
    the productivity of forests. 1. Essential needle-leaved forests of
    Hokkaido. (Shidei, T., ed.) Kokusaku Pulp Co., Tokyo: 73-80**.
Kira, T., Ogawa, H., Yoda, K. and Ogino, K. 1967: Comparative
    ecological studies on three main types of forest vegetation in
    Thailand. 4. Dry matter production with special reference to the
    Khao Chang rain forest. Nature and Life in Southeast Asia 5:
    149-174.
Kirita, H. 1967: (A comparison of litter traps of different shape).
    In: Interim Rept for 1966, Special Project Research "Methods of
    measurements of primary productivity in forests". (Kira, T., ed.):
    65-68***.
Kirita, H., and Hotsumi, K. 1968: (Correction of decrease of weight
    of fallen leaf in litter traps). In: Interim Rept for 1967, Special
    Project Research "Methods of measurements of primary productivity in
    forests". (Kira, T., ed.): 77-80***.
Kishchenko, I.T. 1978: Seasonal growth of pine needles in different
    forest types of Southern Karelia. Lesovedenie No. 2: 29-32. (In
    Russian).
Kittredge, J. 1944: Estimation of the amount of foliage of trees
    and stands. J. For. 42: 905-912.
Kozak, A. 1970: Methods for ensuring additivity of biomass
    components by regression analysis. For. Chron. 46: 1-3.
Krumlik, G.J. and Kimmins, J.P. 1973: Studies of biomass
    distribution and tree form in old virgin forests in the mountains of
    south coastal British Columbia, Canada. In: Young, H.E. (ed.) IUFRO
    Biomass Studies. Univ. Maine Press, Orono, Me. U.S.A. 361-74.
Kubicek, F. 1977: Organic litter production in the oak-hornbeam
    ecosystem. Biologike Prace, 23(1): 131-219.
Lange, O.L., and Schulze, E.D. 1971: Measurement of CO_2
    gas-exchange and transpiration in the beech (Fagus silvatica L.).
    In: Ellenberg, H. (ed.). Integrated experimental ecology.
    Ecological Studies 2. Springer-Verlag, Berlin: 16-28.
Lieth, H. 1972: Modelling the primary productivity of the world.
    Ciencia e Cultra 27: 621-5.

Loomis, R.M., Phares, R.E. and Crosby, J.S. 1966: Estimating foliage and branchwood quantities in shortleaf pine. For. Sci. 12: 30-9.

Madgwick, H.A.I. 1968: Seasonal changes in biomass and annual production of an old-field Pinus virginiana stand. Ecology 49: 149-152.

Madgwick, H.A.I. 1970a: Caloric values of Pinus virginiana as affected by time of sampling, tree age and position in stand. Ecology 51: 1094-7.

Madgwick, H.A.I. 1970b: Biomass and productivity models of forest canopies. In: Reichle, D.E. (ed.) Ecological Studies. Analysis and Synthesis. Vol. 1., Springer-Verlag, Berlin: 47-54.

Madgwick, H.A.I. 1971: The accuracy and precision of estimates of the dry matter in stems, branches and foliage in an old-field Pinus virginiana stand. In: Young, H.E. (ed.) Forest Biomass Studies, Univ. Maine Press, Orono, Me., U.S.A. 105-12.

Madgwick, H.A.I. 1974: Modelling canopy development in pines. In Fries, J. (ed.). Growth Models for Tree and Stand Simulation. Royal College of Forestry, Dept. of Forest Yield Research, Stockholm, Sweden, Res. Notes No. 30: 253-9.

Madgwick, H.A.I. 1975: Branch growth of Pinus resinosa Ait. with particular reference to potassium nutrition. Can. J. For. Res. 5: 509-14.

Madgwick, H.A.I. 1979: Estimating component weights of Pinus radiata. In: Frayer, W.E. (ed.). Forest Resource inventories, Colorado State Univ. Fort Collins Colo., U.S.A.: 717-24.

Madgwick, H.A.I. 1981: Estimating the above-ground weight of forest plots using the basal area ratio method. N.Z. J. For. Sci. (in press).

Madgwick, H.A.I., Beets, P. and Gallagher, S. 1981: Dry matter accumulation, nutrient and energy content of the above ground portion of 4-year-old stands of Eucalyptus nitens and E. fastigata. N.Z. J. For. Sci. 11: 53-9.

Madgwick, H.A.I. and Jackson, D.S. 1974: Estimating crown weights of Pinus radiata from branch variables. N.Z. J. For. Sci. 4: 520-8.

Madgwick, H.A.I., Jackson, D.S. and Knight, P.J. 1977: Above-ground dry matter, energy, and nutrient contents of trees in an age series of Pinus radiata plantations. N.Z. Jour. For. Sci. 7: 445-68.

Madgwick, H.A.I. and Kreh, R.E. 1980: Biomass estimation for Virginia pine trees and stands. For. Sci. 26: 107-11.

Madgwick, H.A.I., Olah, F.D. and Burkhart, H.E. 1977: Biomass of open-grown Virginia pine. For. Sci. 23: 89-91.

Madgwick, H.A.I. and Olson, D.F. 1974: Leaf area index and volume growth in thinned stands of Liriodendron tulipifera L. J. appl. Ecol. 11: 575-80.

Madgwick, H.A.I. and Satoo, T. 1975: On estimating the above-ground weights of tree stands. Ecology 56: 1446-50.

Madgwick, H.A.I., White, E.H., Xydias, G.K. and Leaf, A.L. 1970: Biomass of Pinus resinosa in relation to potassium nutrition. For. Sci. 16: 154-9.

Marks, P.L. 1974: The role of pin cherry (Prunus pensylvanica L.) in the maintenance of stability in northern hardwood ecosystems. Ecological Monographs 44: 73-88.

Marks, P.L. and Bormann, F.H. 1972: Revegetation following forest
    cutting: Mechanisms for return to steady-state nutrient cycling.
    Science 176: 914-5.
Martin, W.L., Sharik, T.L., Oderwald, R.G. and Smith, D.W. 1980:
    Evaluation of ranked set sampling for estimating shrub phytomass in
    Appalachian oak forests. School of Forestry and Wildlife Resources,
    V.P.I. and S.U., Pub. No. FWS4-80, 9 p.
Maruyama, K. 1971: Effect of altitude on dry matter production of
    primeval Japanese beech forest communities in Naeba Mountains. Mem.
    Fac. Agr. Niigata Univ. 9, 85-171.
Maruyama, K., Yamada, M. and Nakazawa, T. 1968: (Tentative
    calculation of photosynthetic gross production of beech natural
    forests: Ecological studies of beech forests 17). Trans. 79th Mtg.,
    Jap. For. Soc., (1968). 286-288**.
Matthews, J.A., Feret, P.P., Madgwick, H.A.I. and Bramlett, D.L.
    1975: Genetic control of dry matter distribution in twenty half-sib
    families of Virginia pine. Proc. 13th Southern Forest Tree
    Improvement Corp, Raleigh N.C. 234-41.
McLaughlin, S.B. and Madgwick, H.A.I. 1968: The effects of position
    in crown on the morphology of needles of loblolly pine (Pinus taeda
    L.). Amer. Midl. Nat. 80: 547-50.
Meyer, H.A. 1938: The standard error of estimates of tree volume
    from the logarithmic volume equation. J. Forestry 36: 340-2.
Meyer, H.A. 1941: A correction for a systematic error occurring
    in the application of the logarithmic volume equation. Penn. State
    Forestry School, Res. Pap. No. 7, 3 p.
Miller, H.G. and Miller, J.D. 1976: Effect of nitrogen supply
    on net primary production in Corsican pine. J. appl. Ecol. 13:
    249-256.
Möller, C.M. 1945: Untersuchungen über Laubmenge, Stoffverlust und
    Stoffprodution des Waldes. Forstl. Forsogsv. Danmark 17: 1-287.
Möller, C.M., Müller, D. and Nielsen, J. 1954a: Respiration in stem
    and branches of beech. Forstl. Forsogsv. Danmark 21: 273-301.
Möller, C.M., Müller, D. and Nielsen, J. 1954b: Graphic
    presentation of dry matter production of European beech. Forstl.
    Forsogsv. Danmark 21: 327-35.
Monsi, M. 1960: Dry matter reproduction in plants. 1. Schemata of
    dry matter reproduction. Bot. Mag. Tokyo 73: 81-90.
Monsi, M., and Saeki, T. 1953: Uber den Lichtfaktor in den
    Pflanzengesellschaften und seine Bedeutung fur die Stoffproduction.
    Jap. J. Bot. 14: 22-52.
Mori, M., et al. 1969: Studies of tending of Akamatsu (Pinus
    densiflora) stands in Tohoku district. 4. Course and present states
    of the experimental plots at Kesennuma. Ann. Rept. Tohoku Branch,
    Govt. For. Expt. Sta., 1968: 211-26**.
Morikawa, Y. 1971: Daily transpiration of a 14-year-old
    Chamaecyparis obtusa stand. J. Jap. For. Soc. 53: 337-339.
Morrison, I.K. and Foster, N.W. 1977: Fate of urea fertilizer added
    to a boreal forest Pinus banksiana Lamb. stand. J. Soil Sci. Soc.
    Amer. 41: 441-8.
Mountford, M.D. and Bunce, R.G.H. 1973: Regression sampling
    with allometrically related variables, with particular reference to
    production studies. Forestry, 46: 203-12.

143

Müller, D., and Nielsen, J. 1965: Production brute, pertes par
respiration et production nette dans la foret ombrophile tropicale.
Forstl. Forsogsv. Danmark 29: 69-160.
Myakushko, V.K. 1974: Reserve of pine forest phytomass according
to zones in the territory of the USSR. Ukrainskii Botanichnii
Zhurnal 31: 205-12 (In Ukranian).
Neenan, M. 1980: The fuel value of coppice wood. In: Production
of Energy from Short Rotation Forestry. An Foras Talúntais, Dublin,
Ireland: 83-6.
Negisi, K. 1970: Respiration in non-photosynthetic organs of trees
in relation to dry matter production of forests. J. Jap. For. Soc.
52: 331-345.
Negisi, K. 1977: Respiration in forest trees. In: Shidei, T. and
Kira, T. (eds.) Primary Productivity of Japanese Forests -
Productivity of Terrestrial Communities, JIBP Synthesis, Tokyo 16:
86-99.
Nemeth, J.C. 1972: Dry matter production and site factors in young
lablolly (Pinus taeda L.) and slash pine (Pinus elliottii Engelm.)
plantations. Ph. D. thesis, North Carolina State Univ., Raliegh,
N.C., U.S.A. 95 p.
Newbould, P.J. 1967: Methods for estimating the primary production
of forests. IBP Handbook 2. Blackwell, Oxford, England, 62 p.
Newhouse, M.E. 1968: Some physiological factors affecting seedling
growth of hardwoods. Ph. D. thesis, Virginia Polytechnic Institute,
Blacksburg, Va. U.S.A., 73 p.
Newhouse, M.E. and Madgwick, H.A.I. 1968: Comparative seedling
growth of four hardwood species. For. Sci. 14: 27-30.
Nomoto, N. 1964: Primary productivity of beech forests in Japan.
Jap. J. Bot. 18: 385-421.
Nordmeyer, A.H. 1980: Phytomass in different tree stands near
timberline. In: Beneke, U. and Davis, M.R. (eds.). Mountain
Environments and Subalpine Tree Growth. N.Z. Forest Service, F.R.I.,
Tech. Pap. No. 70: 111-24.
Ogawa, H., Yoda, K. and Kira, T. 1961: A preliminary survey on the
vegetation of Thailand. Nature and Life in S.E. Asia 1: 21-157.
Ogawa, H., Yoda, K., Ogino, K. and Kira, T. 1965: Comparative
ecological studies on three main types of forest vegetation in
Thailand. 2. Plant biomass. Nature and Life of Southeast Asia 4:
49-80.
Ogino, K., and Shidei, T. 1967: (Biomass and production in beech
forests in Asyu). In: Interim Rept for 1966, Special Research
Project "Methods of measurements of primary productivity in forests"
(Kira, T., ed.): 12-25***.
Ohmann, L.F., Grigal, D.F. and Rogers, L.L. 1980: Estimating plant
biomass for undergrowth species of northeastern Minnesota forest
communities. USDA, For. Serv., Gen. Tech. Rep. NC-61, 10 p.
Olson, D.F. 1971: Sampling leaf biomass in even-aged stands of
yellow-poplar (Liriodendron tulipifera L.). In: Young, H.E. (ed.).
Forest Biomass Studies. Univ. Maine Press, Orono, Me. U.S.A. 115-22.
Oshima, Y., Kimura, M., Iwaki, H. and Kuroiwa, S. 1958: Ecological and
physiological studies on the vegetation of Mt Shimagare. 1.
Preliminary survey of vegetation of Mt Shimagare. Bot. Mag. Tokyo
71: 289-301.

Ovington, J.D. 1953: Studies in the development of woodland conditions under different trees 1. Soils pH. J. Ecol. 41: 13-34.

Ovington, J.D. 1956: The form, weights and productivity of tree species grown in close stands. New Phytol. 55: 289-304.

Ovington, J.D. 1957: Dry matter production by Pinus sylvestris L. Ann. Bot. n.s. 21: 287-314.

Ovington, J.D. 1961: Some aspects of energy flow in plantations of Pinus sylvestris L. Ann. Bot. n.s. 25: 12-20.

Ovington, J.D. 1962: Quantitative ecology and the woodland ecosystem concept. Advances. Ecological Research, 1: 103-92.

Ovington, J.D. 1963: Flower and seed production. A source of error in estimating woodland production, energy flow and mineral cycling. Oikos 14: 148-53.

Ovington, J.D. 1965: Organic production, turnover and mineral cycling in woodlands. Biol. Rev. 40: 295-336.

Ovington, J.D., Forrest, W.G. and Armstrong, J.E. 1967: Tree biomass estimation. In: Young, H.E. (ed.) Symposium on Primary Productivity and Mineral Cycling in Natural Ecosystems. Univ. Maine Press, Orono, Me., U.S.A. 4-31.

Ovington, J.D., and Heitkamp, D. 1960: The accumulation of energy in forest plantations in Britain. J. Ecol. 48: 639-46.

Ovington, J.D., Heitkamp, D. and Lawrence, D.B. 1963: Plant biomass and productivity of prairie, savanna, oakwood and maize field ecosystems in central Minnesota. Ecology 44: 52-63.

Ovington, J.D. and Madgwick, H.A.I. 1959a: The growth and composition of natural stands of birch. Plant and Soil 10: 271-83.

Ovington, J.D. and Madgwick, H.A.I. 1959b: Distribution of organic matter and plant nutrients in a plantation of Scots pine. For. Sci. 5: 344-55.

Ovington, J.D. and Murray, G. 1964: Determination of acorn fall. Quart. J. For. 58: 152-159.

Paterson, S.S. 1956: The forest area of the world and its potential productivity. Medd. f. Univ. Goteborg, Dept. Geogr. No. 51, 216 p + maps.

Pope, P.E. 1979: The effect of genotype on biomass and nutrient content in 11-year-old loblolly pine plantations. Can. J. For. Res. 9: 224-230.

Prodan, M. 1965: Holzmesslehre. Sauerländer's Verlag, Frankfurt a.M., 664 p.

Ranger, J. 1978: Recherches sur les biomasses comparées de deux plantations de Pin laricio de Corse avec ou sans fertilisation. Ann. Sci. Forest., 35: 93-115.

Reineke, L.H. 1933: Perfecting a stand-density index for evenaged forests. J. Agr. Res. 46: 627-638.

Reiners, W.A. 1974: Foliage production by Thuja occidentalis L. from biomass and litter fall estimates. Amer. Midl. Nat. 92: 340-5.

Remezov, N.P. 1959: Method of studying the biological cycle of elements in forest. Soviet Soil Science No. 1. 59-67.

Rennie, P.J. 1955: The uptake of nutrients by mature forest growth. Plant and Soil 7: 49-95.

Riedacker, A. 1968: Methodes indirectes d'estimation de la biomasse des arbres et des peuplements forestiers. Inst. Nat. Rech. Agron. C.N.R.F., 28 p***.

Roberts, J. and Wareing, P.F. 1975: An examination of the differences in dry matter production shown by some progenies of Pinus sylvestris L. Ann. Bot. 39: 311-24.

Rodin, L.E., and Bazilevich, N.I. 1965: Production and mineral cycling in terrestrial vegetation. Translated by G.E. Fogg, 1967: Oliver and Boyd, Edinburgh, London. 288 p.

Rogerson, T.L. 1964: Estimating foliage on loblolly pine. U.S.D.A., For. Serv., Res. Note, SO-16, 3 p.

Rozanov, B.G. and Rozanova, I.M. 1961: The bioligical cycle of nutrient elements of bamboo in the tropical forests of Burma. Bot. Zh. 49.

Rutter, A.J. 1966: Studies on the water relations of Pinus sylvestris in plantation conditions. J. appl. Ecol. 3: 393-405.

Rubtsov, V.I. and Rubtsov, V.V. 1975: Biological production of 20-year Pinus sylvestris (L.) plantations by various density of planting. Lesovedenie No. 1, 28-36 (In Russian).

Safford, L.O. 1974: Effect of fertilisation on biomass and nutrient content of fine roots in a beech-birch-maple stand. Plant and Soil 40: 349-63.

Saito, H., and Shidei, T. 1972: Studies on estimation of leaf fall under model canopy. Bull. Kyoto Univ. For. 43: 162-85*.

Sakaguchi, K. 1961: Studies on basic factors in thinning. Bull. Govt. For. Expt. Sta., Tokyo. 131: 1-95.

Sakaguchi, K., Doi, K. and Ando T. 1955: An analysis of growth of pine (Pinus densiflora) stands in the thicket stage based upon different densities. In: Proc. Japanese Red Pine Symp., Kyoto 1954: 312-27*.

Sakurai, N. 1970: (Effect of fertilisation on Pinus thunbergii forests on coastal sand dune). Thesis, Univ. Tokyo**.

Santantonio, D., Hermann, R.K. and Overton, W.S. 1977: Root biomass studies in forest ecosystems. Pedobiologia 17: 1-31.

Sasa, T. 1969: (Circulation of matter in natural forests of Abies firma and Tsuga sieboldii). Thesis, Univ. Tokyo*.

Sasa, T., and Satoo, T. 1969: (Notes on the recovery of leaf-fall by traps). Trans. 79th Mtg., Jap. For. Soc., (1968): 94-96.**

Sasa, T., and Satoo, T. 1972: (Amount, growth, and mortality of roots of fir and hemlock trees.) In: Interim Rept. for 1971, Special Project Research "Comparative studies of primary productivity of forest ecosystems" (Shidei, T., ed.) 44-52.***

Satoo, T. 1952: (Silviculture). Asakura, Tokyo, 87 p.**

Satoo, T. 1955. (Physical basis of growth of forest trees). In: (Recent advances in silvicultural sciences.) Asakura, Tokyo. 116-41.**

Satoo, T. 1962a. Notes on Kittredge's method of estimation of amount of leaves of forest stands. J. Jap. For. Soc. 44: 267-272.

Satoo, T. 1962b. Notes on Reineke's formulation of the relation between average diameter and density of stands. J. Jap. For. Soc. 44: 343-349.

Satoo, T. 1965: Further notes on the method of estimation of the amount of leaves of forest stand. J. Jap. For. Soc. 47: 185-9.

Satoo, T. 1966. Production and distribution of dry matter in forest ecosystems. Misc. Inform. Tokyo Univ. For. 16, 1-15.

Satoo, T. 1968a. Primary production and distribution of produced dry matter in plantation of Cinnamomum camphora: Materials for the studies of growth in stands. 7. Bull. Tokyo Univ. For. 64: 241-275.

146

Satoo, T. 1968b. Primary production relations in woodlands of <u>Pinus</u>
   <u>densiflora</u>. In: Young, H.E. (ed.). Symposium on Primary
   Productivity and Mineral Cycling in Natural Ecosystems. Univ. Maine
   Press, Orono, Me., U.S.A.: 52-80.
Satoo, T. 1970a. Primary production in a plantation of Japanese
   larch <u>Larix leptolepis</u>: a summarized report of JPTF-66 KOIWAI. J.
   Jap. For. Soc. 52: 154-158.
Satoo, T. 1970b. A synthesis of studies by the harvest method:
   'Primary production relations in the temperate deciduous forests of
   Japan. In: Reichle, D.E. (ed.). Analysis of temperate forest
   ecosystems. Ecological Studies 1. Springer Verlag, Berlin: 55-72.
Satoo, T. (ed.) 1970c. Interim Report for 1969, Special Project
   Research "Primary production in managed forests." pp. 43***
Satoo, T. 1971a. Primary production relations in plantations of
   Norway spruce in Japan: Materials for the studies of growth in
   stands 8. Bull. Tokyo Univ. For. 65, 125-142.
Satoo, T. 1971b. Primary production relations of coniferous forests
   in Japan. In: Productivity of forest ecosystems, Proc. Brussels
   Symp., 1969. (Duvigneaud, P., ed.) UNESCO, Paris. pp. 191-205.
Satoo, T. 1974a. Primary production relations in a natural forest of
   <u>Betula maximowicziana</u> in Hokkaido: Materials for the studies of
   growth in forest stands. 9. Bull. Tokyo Univ. For. 66: 109-7.
Satoo, T. 1974b. Primary production relations in a plantation of
   <u>Larix leptolepis</u> in Hokkaido: Materials for the studies of growth in
   forest stands. 10. Bull. Tokyo Univ. For. 66: 119-26.
Satoo, T. 1974c. Primary production relations in a young plantation
   of <u>Abies sachalinensis</u> in Hokkaido: Materials for the studies of
   growth in forest stands. 11. Bull. Tokyo Univ. For. 66: 127-37.
Satoo, T. 1974d: Primary production relations of a young stand of
   <u>Metasequoia glyptostroboides</u> planted in Tokyo: Materials for the
   studies of growth in forest stands. 13. Bull. Tokyo Univ. For. 66:
   153-64.
Satoo, T. 1979a: Loss of canopy biomass due to thinning - a
   comparison of two young stands <u>of Cryptomeria japonica</u> of cutting and
   seedling origins. J. Jap. For. Soc. 61: 83-7
Satoo, T. 1979b: Standing crop and increment of bole in plantations
   of <u>Chamaecyparis obtusa</u> near an electric power plant in Owase, Mie.
   Jap. J. Ecol. 29: 103-9.
Satoo, T. 1979c: Production of reproductive organs in plantations
   of <u>Chamaecyparis obtusa</u> near an electric power plant in Owase, Mie.
   Jap. J. Ecol. 29: 315-21.
Satoo, T., Kunugi, R. and Kumekawa, A. 1956: Amount of leaves and
   production of wood in an aspen (<u>Populus davidiana)</u> second growth in
   Hokkaido. Materials for the studies of growth in forest stands. 3.
   Bull. Tokyo Univ. For. 52: 33-51.*
Satoo, T., Namakura, K. and Senda, M. 1955: Materials for the
   studies of growth in forest stands. 1. Young stands of Japanese red
   pine of various density. Bull. Tokyo Univ. For. 48: 65-90.*
Satoo, T., Negisi, K. and Senda, M. 1959: Amount of leaves and
   growth in plantations of <u>Zelkowa serrata</u> applied with crown
   thinning: Materials for the studies of growth in forest stands. 5.
   Bull. Tokyo Univ. For. 55: 101-23.*

147

Satoo, T., Negisi, K. and Yagi, K. 1974: Primary production
relations in plantations of <u>Thujopsis dolabrata</u> in the Noto
Peninsula: Materials for the studies of growth in forest stands. 12.
Bull. Tokyo Univ. For. 66: 139-51.

Satoo, T. and Senda, M. 1958: Amount of leaves and production of
wood in a young plantation of <u>Chamaecyparis obtusa</u>: Materials for
the studies of growth in forest stands. 4. Bull. Tokyo Univ. For. 54:
71-100.*

Satoo, T. and Senda, M. 1966: Biomass, dry matter production and
efficiency of leaves in a young <u>Cryptomeria</u> plantation: Materials
for the studies of growth in forest stands. 6. Bull. Tokyo Univ. For.
62: 117-146.*

Schlesinger, W.H. 1978: Community structure, dynamics and nutrient
cycling in the Okefenokee cyprus swamp-forest. Ecol. Monogr. 48:
43-65.

Schumacher, F.X. 1938: New concepts in forest mensuration. J. For.
36: 847-9.

Senda, M., Nakamura, K., Takahara, S. and Satoo, T. 1952: Some
aspects of growth in stands: an analysis of pine stands of different
spacing. Bull. Tokyo Univ. For. 43: 49-57.*

Senda, M., and Satoo, T. 1956: Materials for the studies of growth
in forest stands. 2. White pine (<u>Pinus strobus</u>) stands of various
densitities in Hokkaido. Bull. Tokyo Univ. For. 52: 15-31.*

Shidei, T. (ed.) 1960: Studies on the productivity of forests. 1.
Essential needle-leaved forests of Hokkaido. Kokusaku Pulp. Co.,
Tokyo, 98p**

Shidei, T. 1963: Productivity of Haimatsu (<u>Pinus pumila</u>) community
growing in alpine zone of Tateyama Range. J. Jap. for. Sci. 45:
169-73.

Shidei, T. (ed.) 1964. Studies on the productivity of forests. 2.
Larch (<u>Larix leptolepis</u> Gord.) forests of Shinshu district. Ringyo
Gizitu Kyokai, Tokyo, 60p **.

Shidei, T. (ed.). 1966: Studies on the productivity of forest. 3.
Productivity of sugi forests. Ringyo Gizitu Kyokai, Tokyo, 63.**

Shimaji, K., and Nagatsuka, Y. 1971: Pursuit of the time sequence of
annual ring formation in Japanese fir (<u>Abies firma</u> Sieb. et Zucc.).
J. Jap. Wood. Res. Soc. 17: 122-8.

Shinozaki, K., Yoda, K. Hozumi, K. and Kira, T. 1964: A
quantitative analysis of plant form: the pipe model theory. Jap. J.
Ecol: 14: 97-105, 133-9.

Siemon, G.R. 1973: Effects of thinning on crown structure, stem
form and wood density of radiata pine. Ph. D. Thesis, australian
Natinal University.

Siemon, G.R., Wood, G.B. and Forrest, W.G. 1976: Effect of
thinning on the distribution and biomass of foliage in the crown of
radiata pine. N.Z. J. For. Sci. 6: 57-66.

Singh, R.P. 1980: Energy dynamics in <u>Eucalyptus tereticornis</u>.
Smith plantations in western Uttar Pradesh. Indian For. 106: 649-58.

Spurr, S.H. 1952: Forest Inventory. The Ronald Press, New York,
U.S.A., 476 p.

Steinbeck, K. and Nwoboshi, L.C. 1980: Rootstock mass of coppiced
<u>Platanus occidentalis</u> as affected by spacing and rotation length.
Forest Sci. 26: 545-7.

Stiell, W.W. 1971: Comparative cone production in young red pine planted at different spacings. Can. For. Serv. Publ. 1306, 8p.

Storey, T.G., Fons, W.L. and Sauer, F.M. 1955: Crown characteristics of several coniferous species. U.S. Forest Serv., Div. Fire Res. AFSWP-416, 95 p.

Storey, T.G. and Pong, W.Y. 1957: Crown characteristics of several hardwood tree species. U.S. Forest Serv., Div. Fire Res. AFSWP-968, 36 p.

Strong, T.F. 1980: Energy values of nine Populus clones U.S.D.A., Forest Service, Res. Note NC-257, 3 p.

Swank, W.T. and Schreuder, H.T. 1974: Comparision of three methods of estimating surface area and biomass for a forest of young eastern white pine. For. Sci. 20: 91-100.

Tadaki, Y. 1966: Some discussion on the leaf biomass of forest stands and trees. Bull. Govt. For. Expt. Sta., Tokyo. 184: 135-59.

Tadaki, Y. 1968: Studies on the production structure of forest. 14. The third report on the primary production of a young stand of Castanopsis cuspidata. J. Jap. For. Soc. 50: 60-65.

Tadaki, Y. 1970: Studies on the production structure of forest. 17. Vertical change of specific leaf area in forest canopy. J. Jap. For. Soc. 52: 263-268.

Tadaki, Y. 1976: Biomass of forests with special reference to the leaf biomass of forests in Japan. J. Jap. For. Sci. 58: 416-423.**

Tadaki, Y. and Hatiya, K. 1968: (Forest ecosystems and their productivity). Ringyo Kagaku Gizitu Sinkosyo, Tokyo, 64p

Tadaki, Y., and Hatiya, K. 1970: (Productive structure of two-storeyed forests of conifer and broadleaved trees in subalpine region). Abstracts, 81st Mtg., Jap. For. Soc. (1970): 211-2.**

Tadaki, Y., Ogata, N., Nagatomo, Y and Yoshida, T. 1966: Studies on production structure of forest. 10. Primary productivity of an unthinned 45-year-old stand of Chamaecyparis obtusa. J. Jap. For. Soc. 48: 387-93.*

Tadaki, Y., and Shidei, T. 1960: Studies on production structure of forest. 1. The seasonal variation of leaf amount and dry matter production of deciduous sapling stand (Ulmus parvifolia). J. Jap. For. Soc. 42: 427-34.*

Takahara, S. 1954: Influence of pruning on the growth of Sugi and Hinoki. Bull. Tokyo Univ. For. 46: 1-95.*

Utkin, A.I., Ifanova, M.G. and Ermolova, L.S. 1981: Primary biological productivity of Scotch pine plantations in Vladimir district. Lesovedenie No. 4: 19-27 (in Russian).

Van Cleve, K., Barney, R. and Schlentner, R. 1981: Evidence of temperature control of production and nutrient cycling in two interior Alaska black spruce ecosystems. Can. J. For. Res. 11: 258-73.

Vanselow, K. 1951: Kronen und Zuwachs der Fichte in gleichaltriger Reinbestanden. Forstw. Cbl. 70: 705-719.

Walter, H. 1951: Standortslehre. (Phytologie III-1), 378-379. Eugen Ulmer, Stuttgart. SS. 525.

Waring, R.H., Newman, K. and Bell, J. 1981: Efficiency of tree crowns and stemwood production at different canopy leaf densities. Forestry, 54: 129-37.

Watson, D.J. 1952: The physiological basis of variation in yield. Adv. Agron. 4: 101-45.

Watson, D.J. 1958: The dependence of net assimilation rate on leaf-area index. Ann. Bot. n.s. 22: 37-54.

Whittaker, R.H. 1966: Forest dimensions and production in the Great Smokey Mountains. Ecol. 47: 103-21.

Whittaker, R.H., Bormann, F.H., Likens, G.E. and Siccama, T.G. 1974: The Hubbard Brook ecosystem study: Forest biomass and production. Ecol. Monogr. 44: 233-54.

Whittaker, R.H. and Niering, W.A. 1975: Vegetation of the Santa Catalina Mountains, Arizona. 5. Biomass, production, and diversity along the elevation gradient. Ecology 56: 771-90.

Will, G.M. and Hodgkiss, P.D. 1977: Influence of nitrogen and phosphorus stresses on the growth and form of radiata pine. N.Z. J. For. Sci. 7: 307-20.

Witkowski, Z. and Kosior, A. 1974: Energetics of larval development of the oak leaf roller moth, Tortrix viridana L. (Lepidoptera, Tortricidae) and an estimate of the energy budget in canterpillar development of other insects feeding on oak leaves. Zaklad Ochrony Przyrody, Studia Naturae A. No. 9: 93-106.

Wolter, K.E. 1968: A new method for marking xylem growth. For. Sci. 14: 102-4

Woodwell, G.M., and Botkin, D.B. 1970: Metabolism of terrestrial ecosystems by gas exchange techniques: The Brookhaven approach. In: Reichle, D.E. (ed.). Analysis of temperate forest ecosystems. Ecological Studies 1. Springer Verlag, Berlin: 73-85.

Woodwell, G.M., and Dykeman, W.R. 1966: Respiration of forest measured by CO_2 accumulation during temperature inversions. Science 154: 1031-4.

Worthington, N.P. 1958: How much Douglas-fir will grow on an area? J. For. 56: 763-4.

Yamakura, T., Saito, H. and Shidei, T. 1972: Production and structure of under-ground part of Hinoki (Chamaecyparis obtusa) stand. 1. Estimation of root production by means of root analysis. J. Jap. For. Soc. 54: 118-25.

Yamaoka, Y. 1958: The total transpiration from forest. Trans. Am Geophys. Union 39: 266-72.

Yoda, K. 1971: (Ecology of forest). Tukizi Syokan, Tokyo, 331p**

Yoda, K. et al. 1963: Self-thinning in overcrowded pure stand under cultivated and natural conditions: Intraspecific competition among higher plants. 11. J. Biol. Osaka City Univ. 14: 107-129.

Yoda, K., Shinozaki, K., Ogawa, H., Hozumi, K. and Kira, T. 1965: Estimation of the total amount of respiration in woody organs of trees and forest communities. J. Biol. Osaka City Univ. 16: 15-25.

Young, H.E. 1973: Biomass variation in apparently homogeneous puckerbrush stands. In: Young, H.E. (ed.). IUFRO Biomass Studies. Univ. Maine, Orono, Me., U.S.A. 195-206.

Young, H.E. 1976: A summary and analysis of weight table studies. In: Young, H.E. (ed.) Oslo Biomass Studies. Univ. Maine Press, Orono, Maine, U.S.A., 251-82.

Zajaczkowski, J. and Lech, A. 1981: The effect of different initial growth space on above ground biomass of Scots pine thicket. Paper presented at the Tokyo IUFRO Congress (In press).

Zavitkovski, J. 1981: Small plots with unplanted plot border can distort data in biomass production studies. Can. J. For. Res. 11: 9-12.

150

Zavitkovski, J. and Newton, M. 1971: Litterfall and litter
    accumulation in red alder stands in Western Oregon. Plant and Soil
    35: 257-68.

* in Japanese with English Summary
** in Japanese
Titles in parenthesis are tentative translation for original titles in
Japanese by the author of this book.
*** mimeo. limited distribution.

SUBJECT INDEX

age 49, 57, 61, 64, 87, 96, 98, 104
allocation of increment 104, 119, 132
altitude 53, 70

biomass
    boles 56, 87
    branches 57, 87
    leaves 62, 87
    reproductive structures 82
    roots 82
    understorey 46
bole
    biomass 56, 87
    increment 36, 95
branches
    biomass 57, 87
    increment 37, 96

calorific values 115
crowns, estimating weights 18

drying temperature 20

efficiency
    photosynthetic 117, 121
    to produce boles 132
energy fixation 115

fertilization 67, 96, 104
flowers and fruits 82, 99

genetic variation 78, 110, 127
geographic variation 77
grazing 32

history 2

increment
    allocation 104, 131
    boles 36, 95
    branches 37, 96
    leaves 38, 96
    measurement 33

    reproductive structures 99
    roots 40, 99

leaf area index 8, 81
leaves
    biomass 62, 87
    efficiency 120
    increment 38, 96
    longevity 80, 130
    maximum biomass 14, 64
litter fall 31

methods
    basal area ratio 22
    grazing 32
    litter fall 31
    mean tree 20
    net production 33
    non-destructive 26
    plot size 16
    regression 23
    respiration 41
    sample trees 17
    understorey 31

overstorey
    biomass 49
    increment 99

photosynthesis 6
plot size 16
pollution 78
production
    gross 1, 41, 111
    net 1, 33, 90

reproductive structures 82, 99
respiration 41, 119
roots
    biomass 82
    increment 40, 99
root-shoot ratio 82

seasonal effects 78, 87
silvicultural system 62, 77
site quality 61, 64, 70, 96, 104
solar radiation 5
specific leaf area 10, 120
stand density 12, 53, 57, 61, 70, 96, 98

thinning. 73

understorey 8, 31, 40, 90

variability 16, 22